ANALOG AND DIGITAL VLSI CIRCUIT DESIGN

ANALOG AND DIGITAL VLSI CIRCUIT DESIGN

By

SARADINDU PANDA

Sr. Lecturer

Narula Institute of Technology

Agarpara, West Bengal

UNIVERSITY SCIENCE PRESS

(An Imprint of Laxmi Publications Pvt. Ltd.)

An ISO 9001:2008 Company

BENGALURU • CHENNAI • COCHIN • GUWAHATI • HYDERABAD

JALANDHAR • KOLKATA • LUCKNOW • MUMBAI • RANCHI • NEW DELHI

BOSTON (USA) • ACCRA (GHANA) • NAIROBI (KENYA)

ANALOG AND DIGITAL VLSI CIRCUIT DESIGN

Printed and bound in India
Typeset at Monu Printographics, Delhi
First Edition: 2015
ISBN 978-93-5138-249-2

Branches

©	Bengaluru	080-26 75 69 30	
©	Chennai	044-24 34 47 26,	24 35 95 07
©	Cochin	0484-237 70 04,	405 13 03
©	Guwahati	0361-254 36 69,	251 38 81
©	Hyderabad	040-27 55 53 83,	27 55 53 93
©	Jalandhar	0181-222 12 72	
©	Kolkata	033-22 27 43 84	
©	Lucknow	0522-220 99 16	
©	Mumbai	022-24 91 54 15,	24 92 78 69
©	Ranchi	0651-220 44 64	

PUBLISHED IN INDIA BY

UNIVERSITY SCIENCE PRESS
(An Imprint of Laxmi Publications Pvt. Ltd.)

An ISO 9001:2008 Company
113, GOLDEN HOUSE, DARYAGANJ,
NEW DELHI - 110002, INDIA
Telephone : 91-11-4353 2500, 4353 2501
Fax : 91-11-2325 2572, 4353 2528
www.laxmipublications.com info@laxmipublications.com

C— 10033/015/03
Printed at: SPPPL-DELHI

CONTENTS

PREFACE

The purpose of **Analog and Digital VLSI Circuit Design** is to provide in a single volume a comprehensive reference work covering the broad spectrum of VLSI technology. It is written and developed for the Electronics and Communication Engineers in WBUT. The goal is to provide the most up-to-date information in IC technology, devices and their models, circuit simulations, amplifiers, logic design, memory, Switched Capacitor Circuits, ASIC. This book is not an all-encompassing digest of everything taught within an Electronics Engineering curriculum on VLSI technology. It is an outcome of my experience in teaching this course. I appreciate my students who have recharged my energy in each semester with their suggestions and criticism.

Chapter 1 describes the Introduction to ASIC Design, Design Strategies: Hierarchy, Regularity, Modularity & Locality and Chip Design Options: Gate Array, Field Programmable Gate Array, PLA, PLD, Standard Cell, Full Custom Design.

The text has been divided into ten chapters.

Chapter 2 describes Fabrication & Layout of CMOS, Fabrication Process Flow: Basic steps CMOS

n-Well Process, Layout & Design Rules, CMOS inverter Layout Design.

Chapter 3 describes MOS Inverter Characteristics including Transfer Characteristics: MOS with resistive load, n-MOSFET Load (Enhancement & Depletion), CMOS inverter and Transient Analysis of CMOS Inverter and Delay analysis.

Chapter 4 describes the Static CMOS logic circuits which include different static VLSI logic design styles and DCVSL logic style.

Chapter 5 describes the dynamic logic design styles including Domino CMOS Logic and NORA Logic.

Chapter 6 describes the Sequential CMOS logic circuits including Behaviour of Bi-stable elements, SR Latch Circuit, Clocked JK Latch/Master-Slave JK.

Chapter 7 describes the Subsystem Design including Adders: Carry ahead adder, Manchester carry chain, Multipliers: Serial-parallel Multiplier, array multiplier and High Density Memory: ROM, Static RAM, Dynamic RAM, SD RAM, Flash Memory.

Chapter 8 describes the Physical Design including Floor Planning Methods: Block Placement & Channel Definition, Global and Channel Routing.

Chapter 9 describes the Analog VLSI design including MOS Switches, Voltage follower, CMOS Current Sources and sink, CMOS Differential Amplifier and CMOS Operational Amplifier.

Chapter 10 describes the Resistor realisation using Switched Capacitor, Switched Capacitor Integrator and Switched Capacitor Filters.

I have tried very hard to eliminate the errors in the book, but realize that few may be slipped through. After completing many readings of the final manuscript, I think that I have corrected most of the major errors and hope that the remaining ones are relatively minor in nature. I apologize in advance for those I have missed.

—Author

ACKNOWLEDGEMENT

During the development of this book, I received invaluable help from many people, especially from my colleagues in Narula Institute of Technology. I have not named these individuals in the text since the list would be very long, however, their help is greatly appreciated.

A number of reviewers were instrumental in improving the quality of the manuscript. Their detailed comments on the first draft were very helpful.

I have listed all the work other than my own in the references as much as possible. However, there are still some materials that I cannot credit to the sources, such as those obtained from the Internet, product datasheets, non-technical magazines, and so on.

The compilation of this book would not have been possible without the dedication and efforts of the Editorial Board of Advisors, the section editors, the publishers, and most of all the contributing authors.

Finally, I would like to express my extreme gratitude to my sister Somerita and my parents for doing excellent proofreading and for enduring the long months while this book was being written. She has been a constant support and help during the writing of this manuscript.

—Author

Chapter 1

INTRODUCTION TO ASIC

1.1 INTRODUCTION

ASIC —Application Specific Integrated Circuits (ASICs) refer to those integrated circuits specifically built for specific tasks.

There are many advantages of ASICs as mentioned below :

- Increased speed
- Lower power consumption
- Lower cost (for mass production)
- Better design security (difficult reverse engineering)
- More compact board design (less complex PCB, less inventory costs).

However, there are some disadvantages :

- Long turnaround time from silicon vendors (several weeks)
- Expensive for low-volume production
- Very high NRE cost (high investment in CAD tools, workstations, and engineering manpower)
- Once committed to silicon the design cannot be changed.

1.2 STRUCTURED DESIGN

Over the years, a number of structured design techniques have been developed to deal with complex hardware and software projects. Not surprisingly, the techniques have a great deal of commonality. Design techniques are classified as hierarchy, modularity, and locality.

Hierarchy

The use of hierarchy or "divide and conquer" involves dividing a system into modules, then repeating this process on each module until the complexity of the submodules is at an appropriately comprehensible level of detail. In the case of predefined modules, the design task involves, use of library code intended for the required function. The notion of "parallel hierarchy" can be used to aggregate descriptions in each of the behavioural, structural, and physical domains that represent a design (parallel hierarchy means a hierarchy, which are not necessarily identical, and used in each domain).

Hierarchy allows the use of virtual components and soft versions of the more conventional packaged IC virtual components placed into a chip design as pieces of code, come with support documentation such as verification scripts. They can be supplied by an independent intellectual property (IP) provider or can be reused from a previous product developed in your organization.

Regularity

With regularity as a guide, the designer attempts to divide the hierarchy into a set of similar building blocks. Regularity can exist at all levels of the design hierarchy. Regularity aids in verification efforts by reducing the number of subcomponents to validate and by allowing formal verification programs to operate more efficiently. Design reuse depends on the principle of regularity to use the same virtual component in multiple places or products.

Modularity

The tenet of modularity states that modules have well-defined functions and interfaces. If modules are "well-formed", the interaction with other modules can be well characterised. The notion of "well-formed" may differ from situation to situation, but a good starting point is the criteria placed on a "well-formed" software subroutine. Modularity helps the designer to clarify the approach to a problem, and also allows a design system to more easily check the attributes of a module as its constructed modules also aids in system-on-chip (SOC) designs where a number of IP sources have to be interfaced to complete a design.

Locality

Locality often means temporal locality or adherency to a clock or timing protocol. This is addressed in chapter 7, where different clocking strategies are examined. One of the central themes of temporal locality is to reference all signals that are specified with required set-up and hold times relative to the clock, and outputs have delays related to the edges of the clock.

1.3 CLASSIFICATION OF ASIC

The following digram depicts the various categories of ASIC.

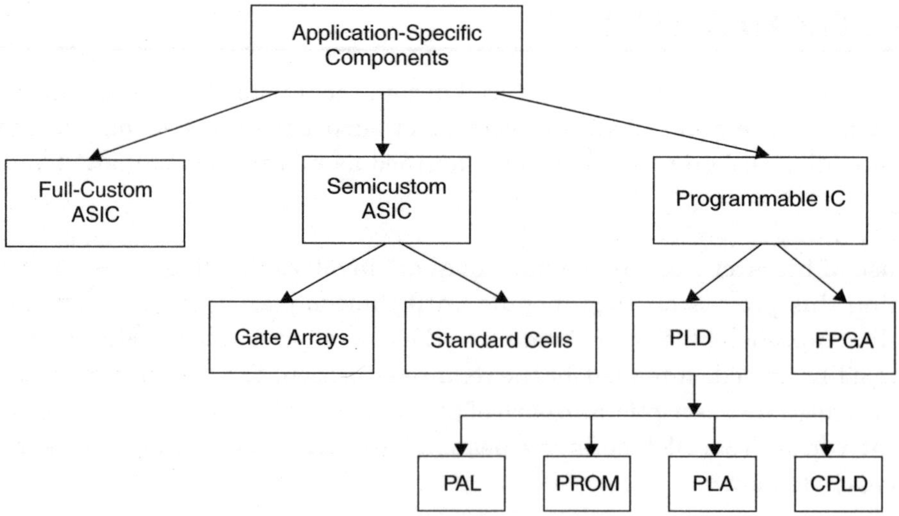

Fig. 1.1. Categories of ASIC

1.3.1 Full-Custom ASIC

- In a full-custom ASIC all mask layers are customized Fig. 1.2(*a*).
- Full-custom designs offer the highest performance and the smallest die size, with the disadvantages of increased design time, higher complexity and costs, together with the highest risk of failure.
- This design option only makes sense when very high performances are required.
- Time after time, fewer projects are really "full-custom," because of the very high cost and the prohibitively slow time-to-market.
- Most of the full-custom works are related to library cell generation or minor parts of a full design.
- Examples of full-custom IC specific parts are high-voltage (automobile, avionic), analog processing and analog/digital communication devices, sensors and transducters.
- Traditionally, microprocessors and memories were exclusively full-custom, but the industry is increasingly turning to semicustom ASIC techniques in these areas too.

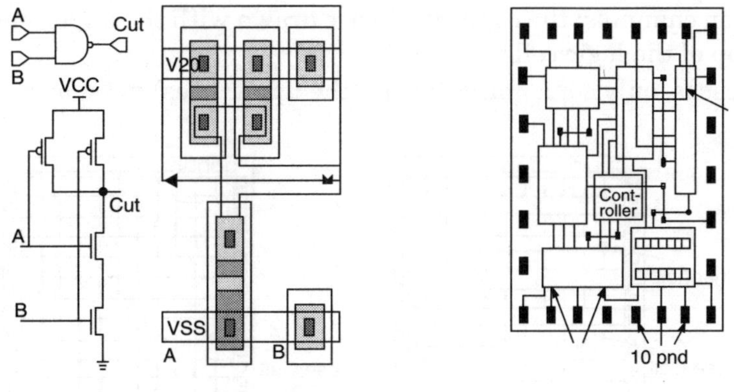

(*a*) In a full-custom design every layer must be defined and Mask layers are customized

(*b*) Design Style

Fig. 1.2. Full-custom ASIC

1.3.2 Semi Custom ASIC

- In order to reduce the unaffordable cost of full-custom in most projects, a wide variety of design approaches have been developed to shorten design time, cut down costs, and automate the processes.
- The approaches are commonly called semi custom.
- Semi custom designs are performed at logic (gate) level.
- As such, they lose some of the flexibility available from a full-custom fashion, that is the price paid for a much easier design technique. Semi custom solutions can be further categorized into **gate array** and **standard cell.**

1.3.3 Gate Array ASIC

- Gate arrays (GAs) are basically composed of continuous arrays of *p*- and *n*-type transistors.
- The silicon vendor provides master or base wafers, to be then personalized according to the interconnection information supplied by the customer.
- Therefore, the designer supplies the personalized information that defines the connections between transistors in the gate array.
- Although a gate array standardizes the chip at the geometry level, user interaction is still typically carried out at logic level.
- The mapping, from transistors to gates, is performed through an ad hoc CAD tool.
- The gate array (also called masked gate array, or prediffused array) uses library components and macros that reduce the development time.
- Two main types of gate arrays can be mentioned: **channeled** and **channelless** (Fig. 1.3(*a*)).
- In a **channeled gate array,** the interconnections are drawn within predefined spaces (channels) between rows of logic cells.
- In a **channelless gate array** (channelless gate array or sea-of gates), there are no connection channels; the connections are drawn with the upper metal layers, that is, on the top of the logic cells.
- In both cases, only some mask layers (the upper ones) must be customized.

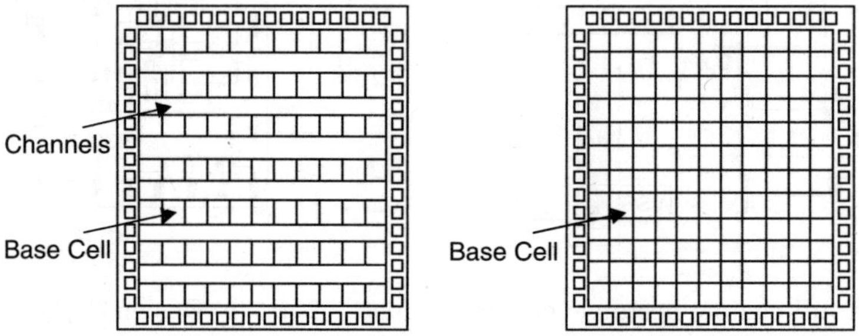

(*a*) Gate array architectures: channeled and channelless gate arrays

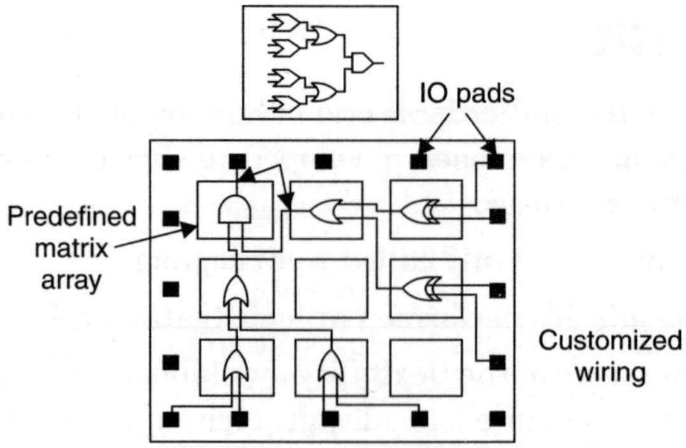

(*b*) Gate array architectures style

Fig. 1.3

1.3.4 Standard Cell Based ASIC

- Standard cells are logic components (*i.e.*, gates, multiplexers, adders, flip-flops) previously designed and stored in a library.
- A design is created using these library cells as inputs to a CAD system: logic schematic diagram or hardware description language (HDL) code description.
- Next, a further CAD tool automatically converts the design into a chip layout.
- Standard-cell designs are typically organized on the chip, as rows of constant height (Fig. 1.4).
- Together with logic-level component cells, standard cell systems typically offer-higher level functions such as multipliers and memory arrays.
- This allows the use of predesigned (or automatically generated) high-level components to complete the design.

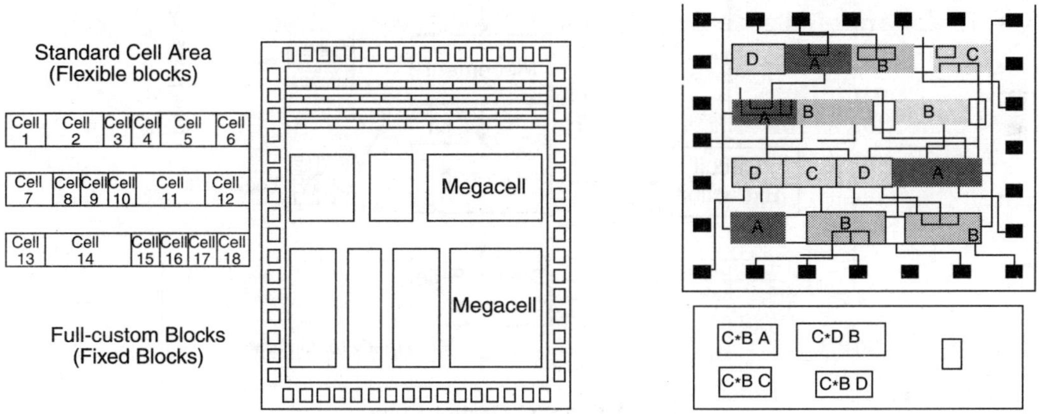

(*a*) Typical standard cell layout (*b*) Typical standard cell design style

Fig. 1.4. Standard cell-based ASIC

1.4 DESIGN FLOW IN ASIC

- Figure 1.5 shows a typical semicustom ASIC design flow (except for the test vector generation).
- ***Design Entry*** enters the design using either a hardware description language such as VHDL or Verilog or a schematic entry.
- ***Logic Synthesis*** from the HDL or schematic entry, extracts a netlist, that is, a description of the logic cells and their connections. The synthesis tool can infer a hardware implementation with the behaviour as the HDL description.
- ***System Partitioning*** divides a large system into ASIC-sized pieces.
- ***Prelayout (Behavioural) Simulation*** checks the circuit working.
- ***Floorplanning*** arranges the different blocks of the circuit on the chip.
- ***Placement*** sets the cell locations in a block.
- ***Routing*** creates the connections between cells and blocks.

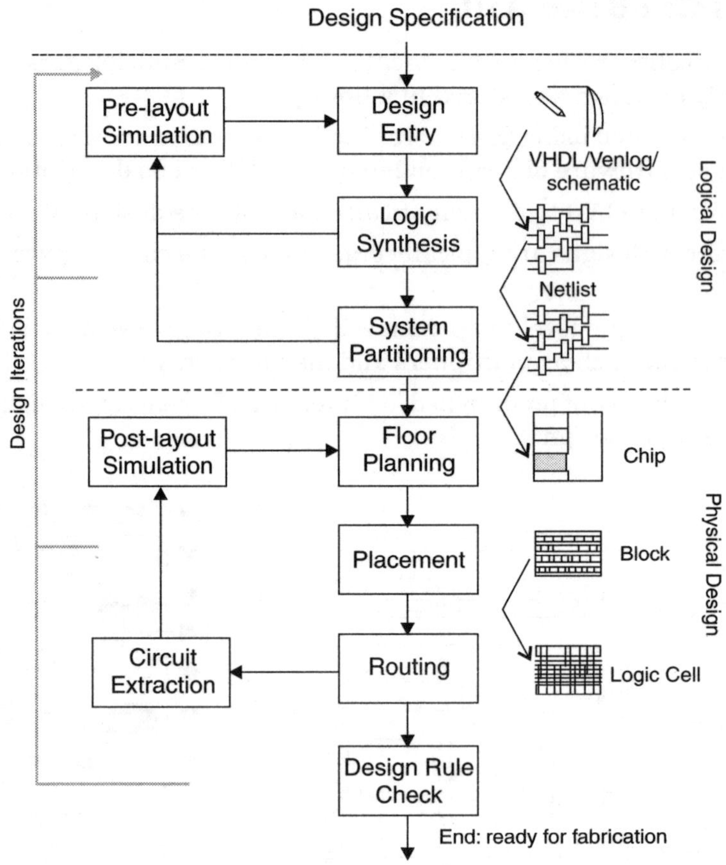

Fig. 1.5. ASIC Design Flow

- ***Extraction (Back Annotation)*** determines the resistance and capacitance of the interconnections and calculates delays for simulation purposes.
- ***Postlayout (Physical) Simulation*** checks the circuit working after including the delays created by interconnection loads.
- ***Design Rule Check (DRC)*** verifies that the circuit layout complies with the specifications of the design rules. DRC tools can range from a simple physical spacing check-up to complex tests.

1.5 PROGRAMMABLE LOGIC DEVICES

- Logic devices can be classified into two broad categories: **fixed** and **programmable.**
- Circuits in a fixed logic device are permanent. They perform one function or set of functions, and once manufactured, they cannot be changed, as traditional ASICs can.
- On the other hand, programmable logic devices (PLDs) are standard, off the-shelf parts that can be modified at any time to perform any number of functions.

A key benefit of using PLDs

(*i*) During the design phase, designers can change circuitry as often as they want until the design operates satisfactorily.

(*ii*) PLDs are based on rewritable memory technology to modify the design, the device only needs to be reprogrammed.

- Reusability is a further attractive feature of PLDs.
- Many types of programmable logic devices are currently available.
- The range of market products includes small devices capable of implementing a handful of equations up to huge FPGAs that can hold an entire processor core plus a number of peripherals.
- Besides this impressive diversity of sizes, numerous alternative architectures are offered to the designer.

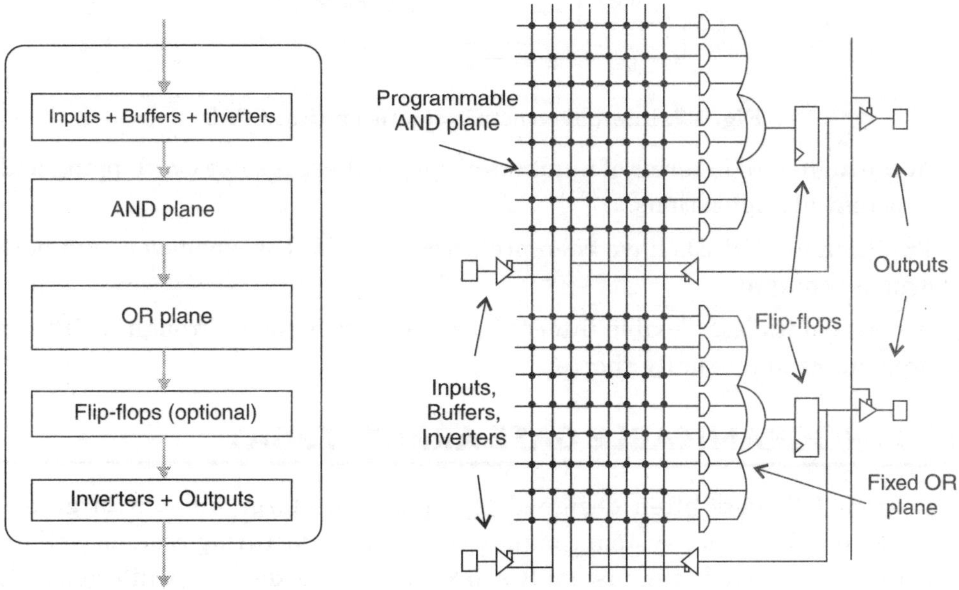

Fig. 1.6. Typical PLD architecture

- Within programmable logic devices, two major types deserve to be highlighed: the ***complex programmable logic device (CPLD)*** and ***field programmable gate array (FPGA)***. These are described below.
- As chip densities increased, PLD manufacturers naturally developed their products toward larger parts, called complex programmable logic devices (CPLDs).

1.6 COMPLEX PROGRAMMABLE LOGIC DEVICE (CPLD)

- In a certain respect, CPLDs can be described as several PLDs (plus some programmable interconnection) in a single chip. The larger size of a CPLD allows implementing either more logic equations or more complicated designs.
- Figure 1.7 contains a block diagram of a typical CPLD (within each logic block stands the equivalent of one PLD).
- Because CPLDs can hold larger designs than PLDs, their potential uses are quite wide-ranging. Sometimes they are used for simple applications, like address decoding,

but more often they contain high performance control-logic or complex finite state machines.

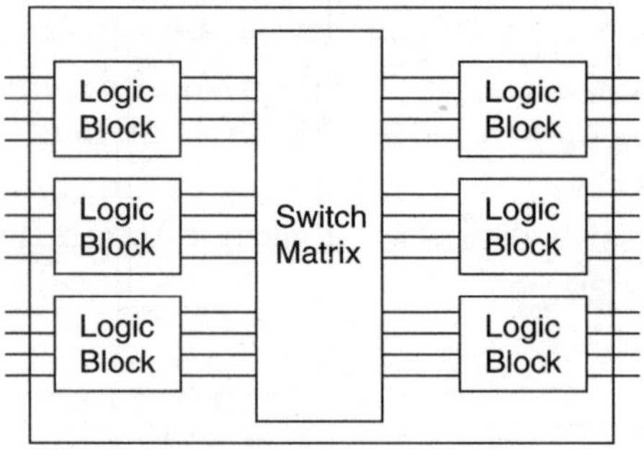

Fig. 1.7. Internal structure of a theoretical CPLD

- At the high-end (in terms of numbers of gates), there is some overlapping with FPGAs in potential applications.
- Traditionally, CPLDs have been preferred over FPGAs whenever high performance logic is required.
- Because of its less flexible internal architecture, delays through a CPLD are more predictable and usually shorter.

1.7 FIELD PROGRAMMABLE GATE ARRAY (FPGA)

- Custom ICs were often designed to replace the large amount of glue logic and consequently reduce system complexity and manufacturing cost, as well as improve performances. However, custom ICs are expensive to develop, while generating time-to-market (TTM) delays because of the prohibitive design time.
- Therefore, the custom IC approach was only viable for products with very high volume (lowering the NRE cost impact), and not TTM sensitive. Coping with this problem, XilinxTM (a startup company) introduced, in 1984, the FPGA technology as an alternative to custom ICs for implementing glue logic.
- The basic FPGA architecture consists of a two dimensional array of logic blocks and flip-flops with means for the user to configure:

  (*i*) the function of each logic blocks,

  (*ii*) the inputs/outputs, and

  (*iii*) the interconnection

  between blocks (Fig. 1.8).
- Families of FPGAs differ from each other by the physical means for implementing user programmability, arrangement of interconnection wires, and basic functionality of the logic blocks.

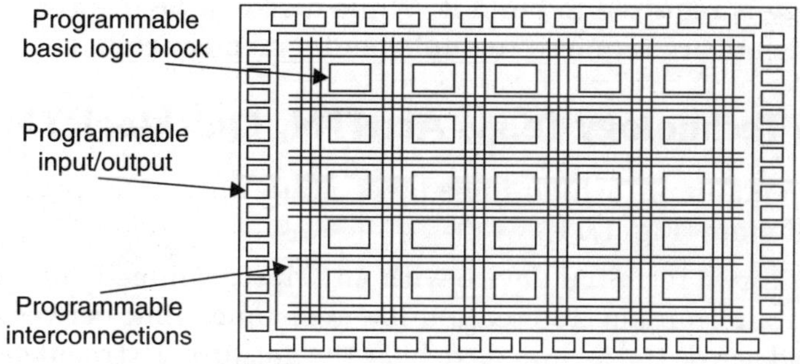

Fig. 1.8. Basic architecture of FPGA: two-dimensional array of progammable logic cells, interconnection, and inout/output

1.8 PROGRAMMING METHODS

There are three main types of programmability as mentioned below :

1.8.1 SRAM Based (e.g., Xilinx™, Altera™)

FPGA connections are achieved using pass-transistors, transmission gates, or multiplexers that are controlled by SRAM cells (Fig. 1.9). This technology allows fast-in-circuit reconfiguration.

The major disadvantages are the size of the chip, required by the RAM technology, and the needs of some external source (usually external nonvolatile memory chips) to loads the chip configuration. The FPGA can be programmed an unlimited number of times.

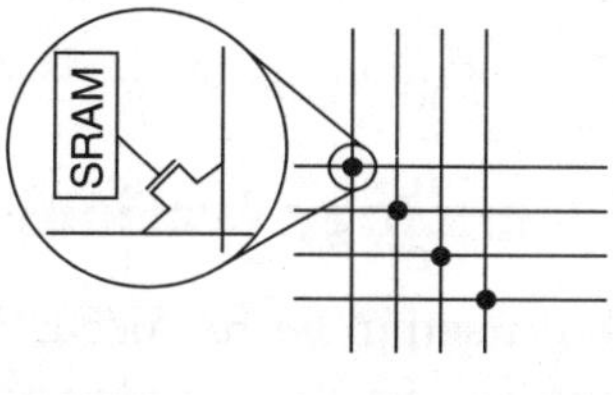

Fig. 1.9. Programming methods: SRAM connection

Static RAM cells are used to control pass gates or multiplexers. To use pass gate as closed switch, boolean one is stored in SRAM cell. When zero is stored, pass transistor provides high resistance between two wire segments. Figure 1.10 (*a*) depicts this usage of SRAM.

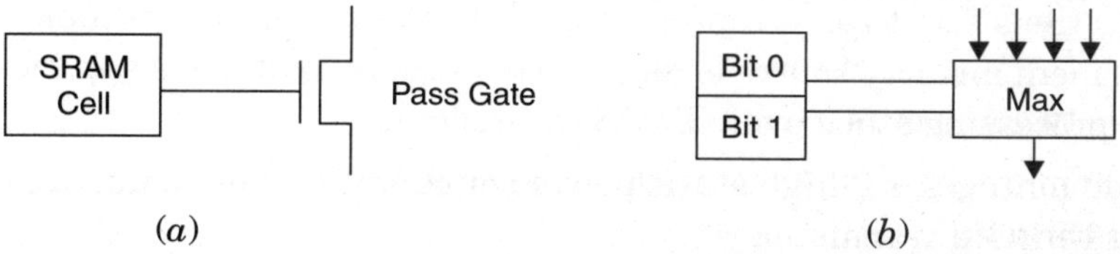

Fig. 1.10. Usage of SRAM

To use SRAM as multiplexer, state of control values stored in SRAM decides which of the multiplexer inputs are connected to the output as shown in Fig. 1.10 (*b*).

Advantage of SRAM is that it provides fast re-programmability and integrated circuit fabrication technology is required to build it. While disadvantage is the space it consumes as minimum five transistors are required to implement a memory cell.

1.8.2 Antifuse Technology (e.g., ActelTM, QuicklogicTM)

An antifuse remains in a high-impedance state until it is programmed into a low-impedance or "fused" state (Fig. 1.11).

An antifuse is a two terminal device with an unprogrammed state providing very high resistance between its terminals. To create a low resistance link between the two terminals high voltage is applied across the terminals to blow the antifuse. Extra bit of circuitry is required to program an antifuse. Antifuse technology is used by FPGA's from Actel, Quick Logic and Crosspoint.

Advantage of antifuse is relatively small size and hence area reduction which is anulled by area consumed by extra circuitry to program it. Another big advantage is low series resistance and low parasitic capacitance.

This technology can be used Only One-Time Programmable (OTP) devices; it is less expensive than the RAM technology.

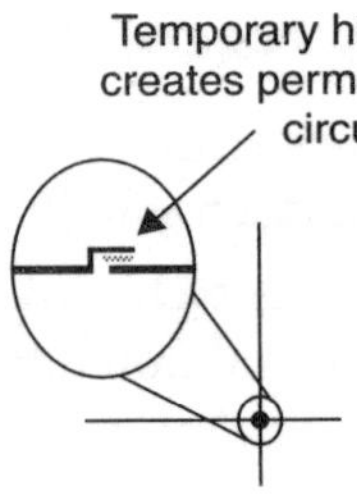

Fig. 1.11. Programming methods: antifuse

1.8.3 EPROM/EEPROM Technology (various PLDs)

This method is the same as that used in EPROM/EEPROM memories. The configuration is stored within the devices, that is, without external memory. Generally, in-circuit reprogramming is not possible.

Technology found in ultraviolet erasable EPROM and electrically erasable EEPROM devices is used in FPGA from Altera. The programmable switch is transistor that permanently be disabled.

Here again the advantage is reprogrammability but there is another advantage that is no external permanent memory source is required to program it at power-up. However, it requires three additional processing steps over CMOS technology.

Other disadvantages are high static power consumption due to pull up resistor and high ON-resistance of EPROM transistor.

Electrically programmable EPROM is used by AMD and Lattice. Use of EEPROM gives advantage of easy reprogrammability.

However, EEPROM cell is twice as large as EPROM cell.

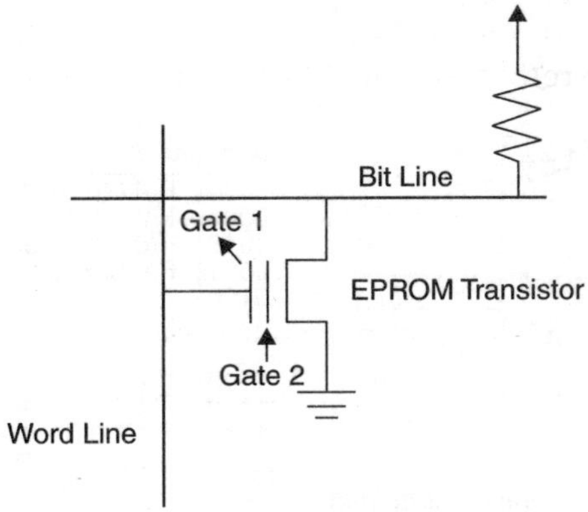

Featuring Gates

Fig. 1.12

1.9 LOOK-UP TABLES

The way logic functions are implemented in a FPGA is another key feature.

Logic blocks that carry out logical functions are look-up tables (LUTs), implemented as memory, or multiplexer and memory.

Figure 1.13 shows these aternatives, together with an example of memory contents for some basic operations.

A $2^n - 1$ ROM can implement any n-bit function. Typical sizes for n are 2, 3, 4 or 5.

In Fig. 1.13 (*a*), an n-bit LUT is implemented as a $2^n - 1$ memory; the input address selects one of 2^n memory locations. The memory locations (latches) are normally loaded with values from user's configuration bit-stream.

In Fig. 1.13 (*b*), the multiplexer control inputs are the LUT inputs. The result is a general-purpose "logic gate." An n-LUT can implement any n-bit function. An n-LUT is a direct implementation of a function truth table.

Each latch location holds the value of the function corresponding to one input combination.

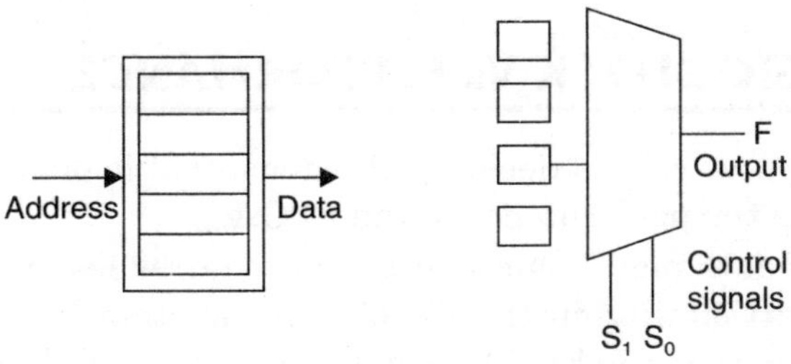

Inputs	AND	OR	XOR	NAND
0 0	0	0	0	1
0 1	0	1	1	1
1 0	0	1	1	1
1 1	1	1	0	0

(*c*)

Fig. 1.13. Look-up table implemented as (*a*) memory or (*b*) multiplexer and memory (*c*) Memory contents example for different logic functions

An example of a 4-LUT is shown in Fig. 1.14.

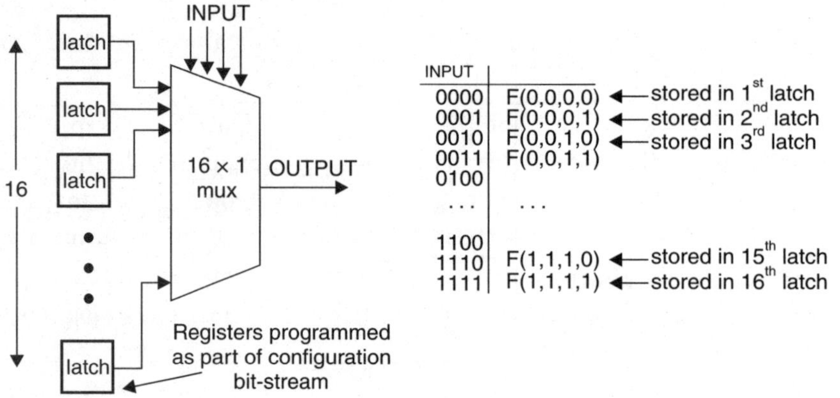

Fig. 1.14. 4-LUT implementation and truth-table contents

1.10 FPGA LOGIC BLOCK

A simplified FPGA logic block can be designed with a LUT, typically a 4-input LUT, implementing a combinational logic function, and register that optionally stores the output of the logic generator (Fig. 1.15).

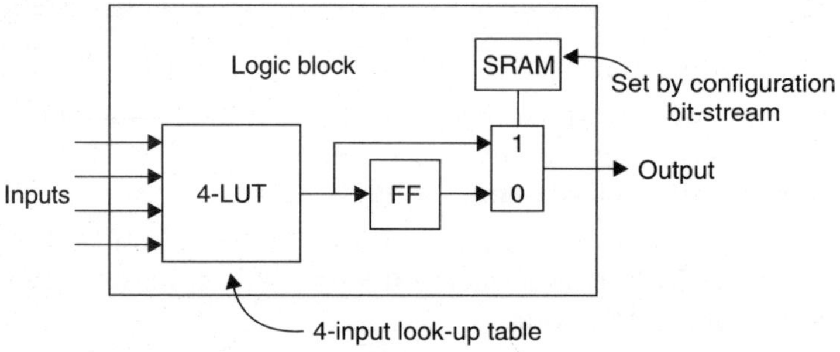

Fig. 1.15. A basic FPGA logic block

1.11 TRADE-OFF SIZE OF LOGIC BLOCK Vs PERFORMANCE

Size of a logic block plays an important role in deciding the density of logic blocks and area utilization in an FPGA. It also affects the performance of the FPGA.

- A large size logic block implements more logic and hence requires less number of logic blocks to implement a functionality on the FPGA. On the other hand a large logic block will consume more space on the FPGA. So optimal size of logic block is one that optimally uses lesser number of logic blocks for functionality implementation while consuming as little space as possible.
- Active logic area is generally less than total logic area due to presence of programmable connections. Total logic area is sum of active logic area and area consumed by programmable connections.

- Routing area of an FPGA is typically more than the active area. It is 70 to 90% of total area of an FPGA.
- In case of Look-up table based FPGA, a 4-input look-up table gives best results in terms of logic synthesised and area consumed.
- Granularity of logic block has influence on performance of an FPGA. Typically higher granularity level results in lesser delay between input and output. As the granularity of logic block increases, number of levels of logic in critical path decreases, and hence delay in critical path decreases. On the flip side with increase in granularity level average fan out increases and number of switches also increases as each block has more pins. Also the length of wires increases with increase in size of logic block.

1.12 FPGA GENERIC DESIGN FLOW

The FPGA design flow has several points in common with the semi custom ASIC design flow.

Figure 1.16 presents a simplified FPGA design flow.

The successive process phases (blocks) of Fig. 1.16 are described as follows:

- **Design Entry:** Creation of design files using schematic editor or hardware description language (Verilog, VHDL, Abel).
- **Design Synthesis:** A process that starts from a high level of logic abstraction (typically Verilog or VHDL) and automatically creates a lower level of logic abstraction using a library of primitives.

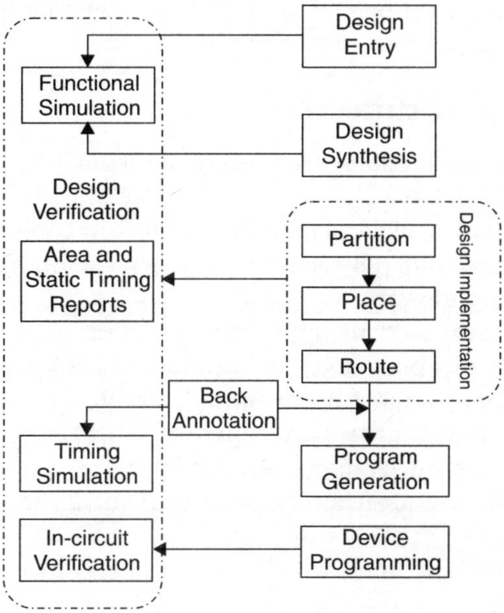

Fig. 1.16. FPGA design flow

- **Partition (or Mapping):** A process assigning to each logic element a specific physical element that actually implements the logic function in a configurable device.
- **Place:** Maps logic into specific locations in the target FPGA chip.
- **Route:** These are connections of the mapped logic.
- **Program Generation:** A bit-stream file is generated to program the device.
- **Device Programming:** Downloading the bit-stream to the FPGA.
- **Design Verification:** Simulation is used to check functionalities. The simulation can be done at different levels. The functional or behavioural simulation does not take into account component or interconnection delays.

  The timing simulation uses back-annotated delay information extracted from the circuit.

  Other reports are generated to verify other implementation results, such as maximum frequency and delay and resource utilization.
- The partition (or mapping), place, and route processes are commonly referred to as **design implementation.**

1.13 FPGA ROUTING TECHNIQUES

- A wire segment can be described as two end points of an interconnect with no programmable switch between them.
- A sequence of one or more wire segments in an FPGA can be termed as a track.
- Typically an FPGA has logic blocks, interconnects and Input/Output blocks.
- Input Output blocks lie in the periphery of logic blocks and interconnect.
- Wire segments connect I/O blocks to wire segments through connection blocks.
- Connection blocks are connected to logic blocks, depending on the design requirement one logic block is connected to another and so on.

1.13.1 Xilinx Routing Architecture

- In Xilinx routing, connections are made from logic block into the channel through a connection block.
- As SRAM technology is used to implement Lookup Tables, connection sites are large.
- A logic block is surrounded by connection blocks on all four sides. They connect logic block pins to wire segments.
- Pass transistors are used to implement connection for output pins, while use of multiplexers for input pins saves the number of SRAM cells required per pin.
- The logic block pins connecting to connection blocks can then be connected to any number of wire segments through switching blocks.

There are four types of wire segments available:

1. **General purpose segments:** the ones that pass through switches in the switch block.
2. **Direct interconnect:** ones which connect logic block pins to four surrounding connecting blocks.
3. **Long line:** high fan out uniform delay connections
4. **Clocks lines:** clock signal provider which runs all over the chip.

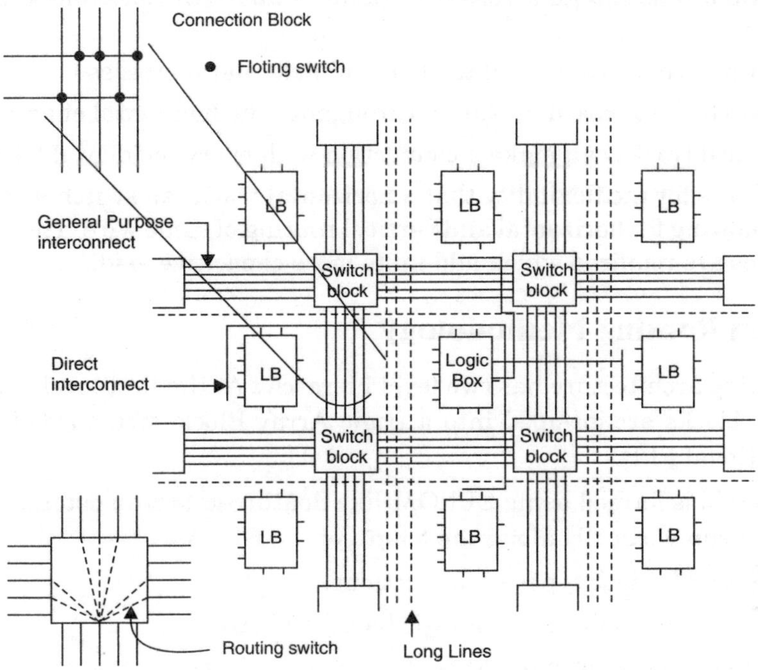

Fig. 1.17. Routing techniques in FPGA

1.13.2 Actel Routing Methodology

- Actel's design has more wire segments in horizontal direction than in vertical direction.
- The input pins connect to all tracks of the channel that is on the same side as the pin.

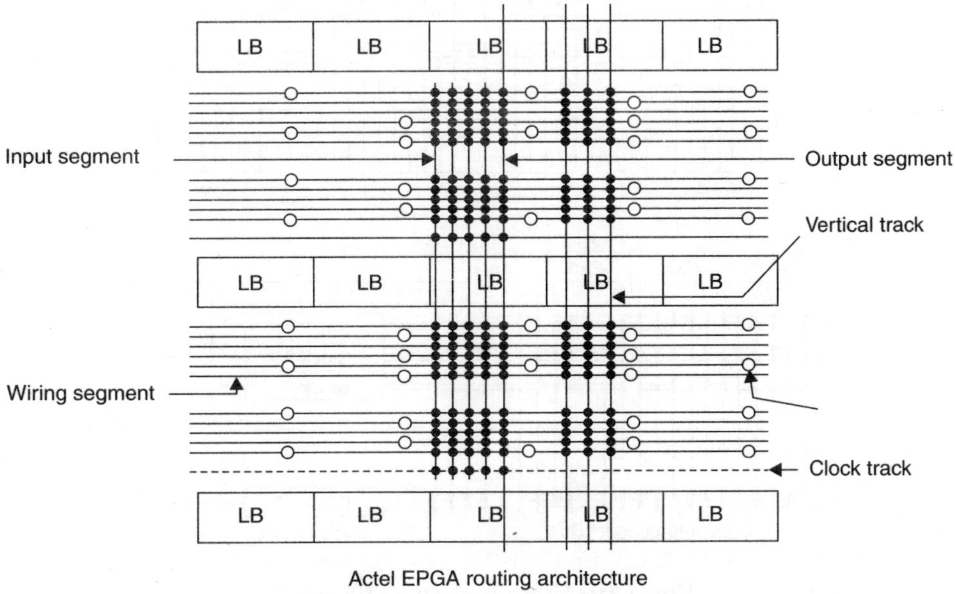

Fig. 1.18. Actel routing methodology

- The output pins extend across two channels above the logic block and two channels below it.
- Output pin can be connected to all 4 channels that it crosses.
- The switch blocks are distributed throughout the horizontal channels.
- All vertical tracks can make a connection with every incidental horizontal track.
- This allows for the flexibility that a horizontal track can switch into a vertical track, thus allowing for horizontal and vertical routing of same wire. The drawback is more switches are required which add up to more capacitive load.

1.13.3 Altera Routing Methodology

Altera routing architecture has two level hierarchy. At the first level of the hierarchy, 16 or 32 of the logic blocks are grouped into a Logic Array Block, structure of the LAB is very similar to a traditional PLD.

The connection is formed using EPROM-like floating-gate transistors. The channel here is set of wires that run vertically along the length of the EPGA.

Tracks are used for four types of connections:

- connections from output of all logic blocks in LAB.
- connection from logic expanders.
- connections from output of logic blocks in other LABs
- connections to and from Input output pads

All four types of tracks connect to every logic block in the array block.

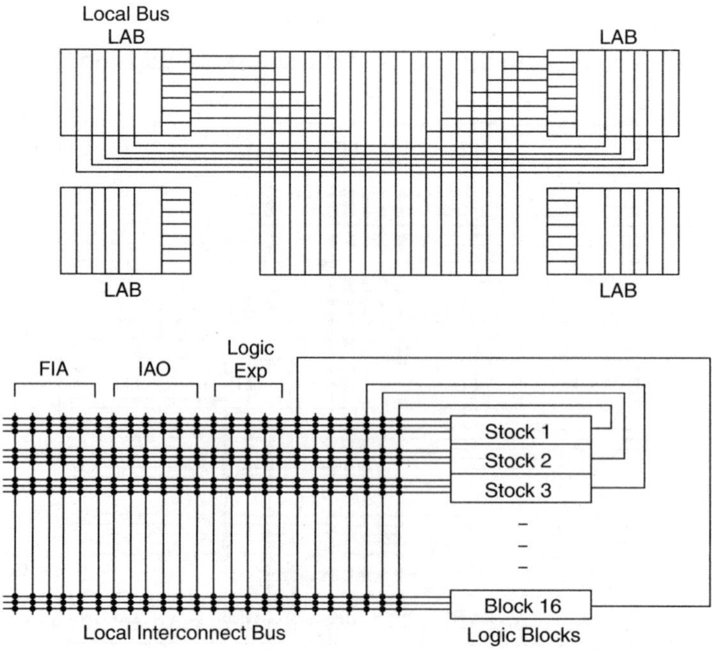

Fig. 1.19. Altera routing methodology

The connection block makes sure that every such track can connect to every logic block pin.

Any track can connect to any input which makes this routing simple.

The intra-LAB routing consists of segmental channel, where segments are as long as possible.

Global interconnect structure called programmable interconnect array (PIA) is used to make connections among LABs.

Its internal structure is similar to internal routing of a LAB.

Advantage of this scheme is that regularity of physical design of silicon allows it to be packed tightly and efficiently.

The disadvantage is the large number of switches required, which adds to capacitive load.

1.14 FPGA STRUCTURAL CLASSIFICATION

Basic structure of an FPGA includes logic elements, programmable interconnects and memory. Arrangement of these blocks is specific to particular manufacturer.

On the basis of internal arrangement of blocks FPGAs can be divided into three classes:

1.14.1 Symmetrical Arrays

This architecture consists of logic elements (called CLBs) arranged in rows and columns of a matrix and interconnect laid out between them.

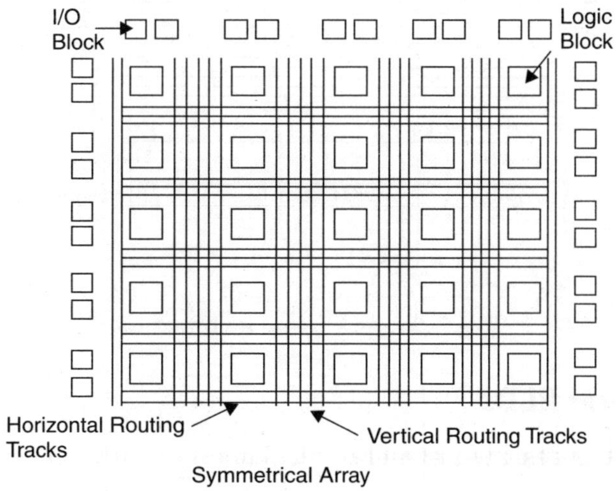

Fig. 1.20. Symmetrical array

This symmetrical matrix is surrounded by I/O blocks which connect it to outside world.

Each CLB consists of n-input Lookup table and a pair of programmable flip flops.

I/O blocks also control functions such as tri-state control, output transition speed.

Interconnects provide routing path.

Direct interconnects between adjacent logic elements have smaller delay compared to general purpose interconnect.

1.14.2 Row Based Architecture

Row based architecture consists of alternating rows of logic modules and programmable interconnect tracks.

Input output blocks are located in the periphery of the rows. One row may be connected to adjacent rows via vertical interconnect.

Logic modules can be implemented in various combinations.

Combinatorial modules contain only combinational elements, but sequential modules contain both combinational elements along with flip flops.

This sequential module can implement complex combinatorial-sequential functions.

Routing tracks are divided into smaller segments connected by antifuse elements between them.

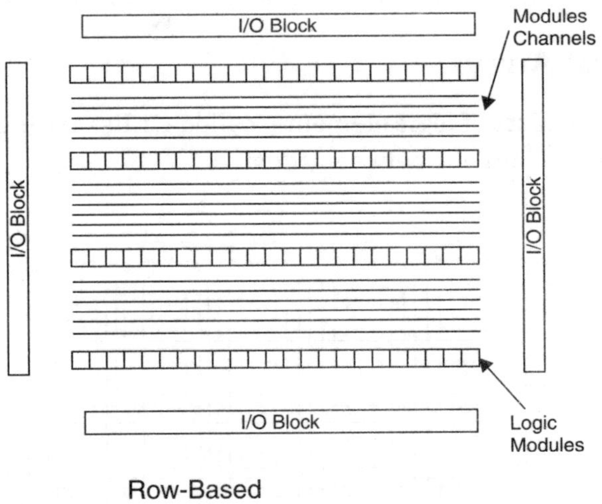

Fig. 1.21. Row based architecture

1.14.3 Hierarchical PLDs

This architecture is designed in hierarchical manner with top level containing only logic blocks and interconnects.

Each logic block contains number of logic modules.

And each logic module has combinatorial as well as sequential functional elements.

Each of these functional elements is controlled by the programmed memory.

Communication between logic blocks is achieved by programmable interconnect arrays. Input output blocks surround this scheme of logic blocks and interconnects.

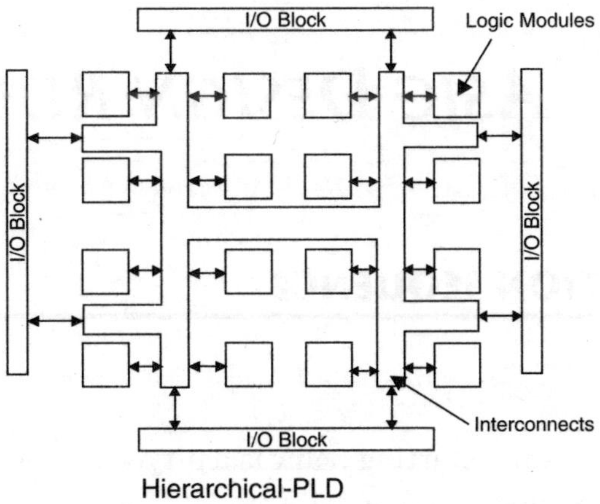

Fig. 1.22. Hierarchical PLD

Chapter 2

ASIC DESIGN RULES

2.1 CMOS FABRICATION SEQUENCE

2.1.1 Start

— For an *n*-well process the starting point is a *p*-type silicon wafer :

— wafer typically 75 to 300 mm in diameter and less than 1 mm thick.

2.1.2 Epitaxial Growth

— A single *p*-type single crystal film is grown on the surface of the wafer by

- subjecting the wafer to high temperature and a source of dopant material

— The epi layer is used as the base layer to build the devices.

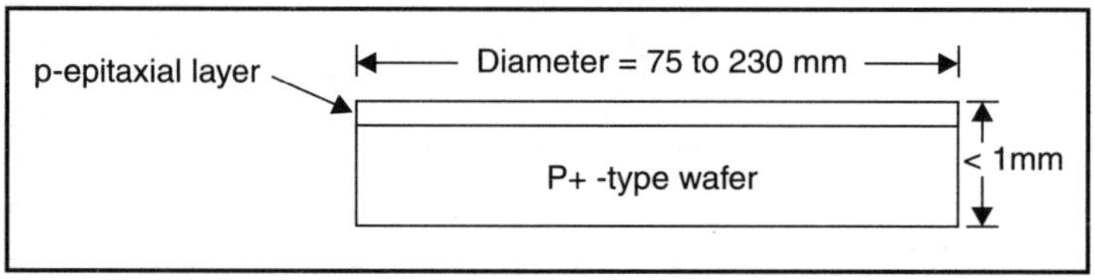

2.1.3 N-well Formation

— PMOS transistors are fabricated in *n*-well regions

— The first mask defines the *n*-well regions

— N-well's are formed by ion implantation or deposition and diffusion

— Lateral diffusion limits the proximity between structures

— Ion implantation results in shallower wells compatible with today's fine-line processes

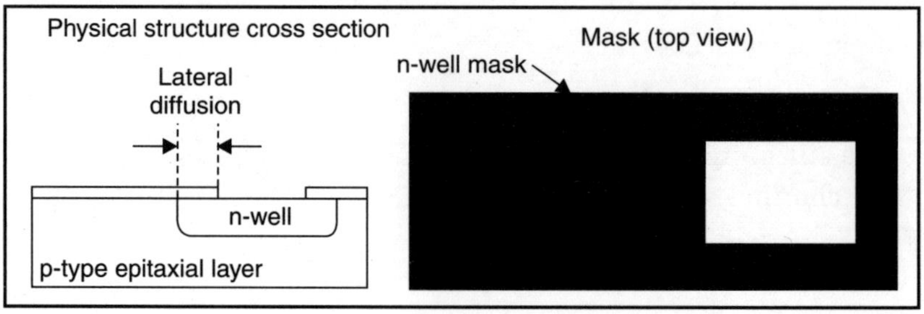

2.1.4 Active Area Definition

— Active area :

- planar section of the surface where transistors are build
- defines the gate region (thin oxide)
- defines the n^+ or p^+ regions

— A thin layer of SiO_2 is grown over the active region and covered with silicon nitride

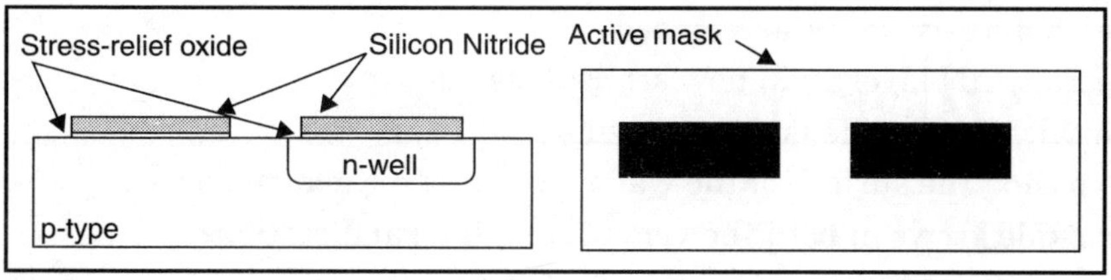

2.1.5 Isolation

— Parasitic (unwanted) FET's exist between unrelated transistors (Field Oxide FET's)

— Source and drains are existing source and drains of wanted devices

— Gates are metal and polysilicon interconnects

— The threshold voltage of FOX FET's are higher than for the normal FET's

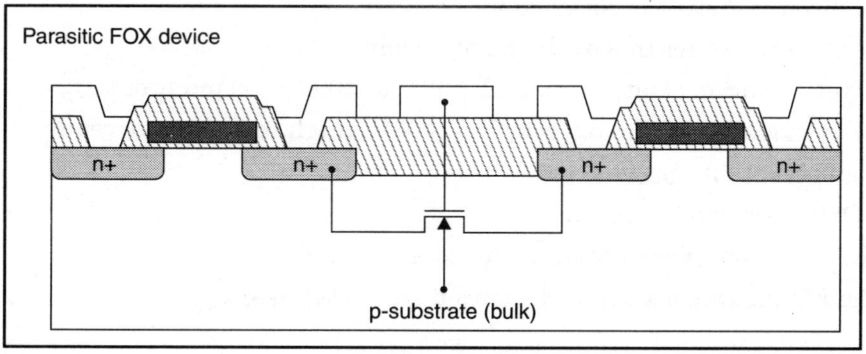

— FOX FET's threshold is made high by

- introducing a channel-stop diffusion that raises the impurity concentration in the substrate in areas where transistors are not required
- making the FOX thick.

2.1.5.1 Channel-stop Implant

— The silicon nitride (over n-active) and the photoresist (over n-well) act as masks for the channel-stop implant.

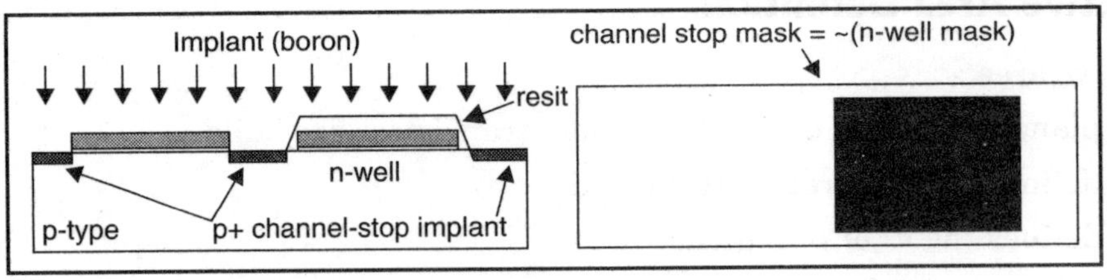

2.1.5.2 *Local Oxidation of Silicon (LOCOS)*

— The photoresist mask is removed

— The SiO_2/SiN layers will now act as a masks

— The thick field oxide is then grown by:

- exposing the surface of the wafer to flow of oxygen-rich gas

— The oxide grows in both the vertical and lateral directions

— This results in a active area smaller than patterned

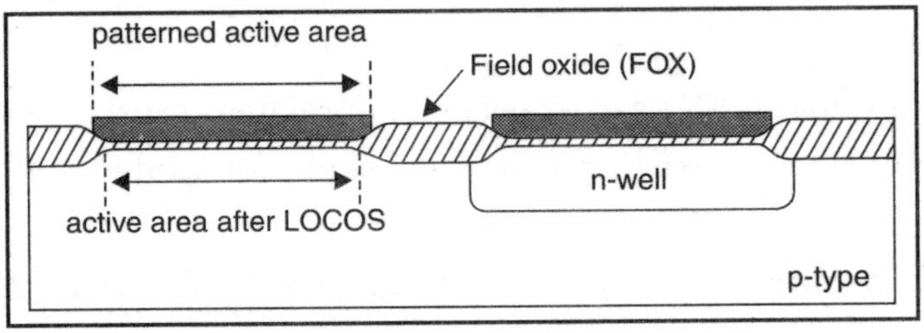

- Silicon oxidation is obtained by:

— Heating the wafer in a oxidizing atmosphere:

- Wet oxidation : water vapour, T = 900 to 1000°C (rapid process)
- Dry oxidation : Pure oxygen, T = 1200°C (high temperature required to achieve an acceptable growth rate)
- Oxidation consumes silicon

— SiO_2 has approximately twice the volume of silicon

— The FOX is recedes below the silicon surface of $0.46X_{FOX}$

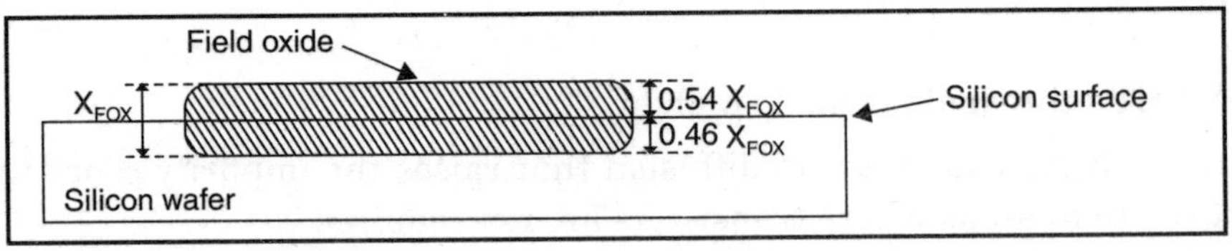

2.1.6 Gate Oxide Growth

— The nitride and stress-relief oxide are removed

— The devices threshold voltage is adjusted by

- adding charge at the silicon/oxide interface

— The well-controlled gate oxide is grown with thickness t_{ox}

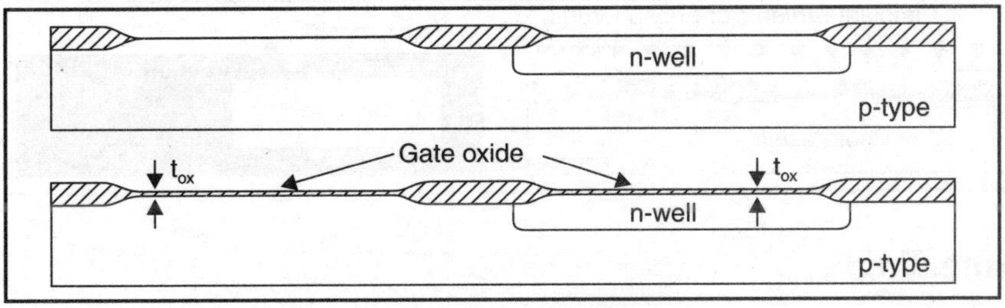

2.1.7 Polysilicon Deposition and Patterning

— A layer of polysilicon is deposited over the entire wafer surface
— The polysilicon is then patterned by a lithography sequence
— All the MOSFET gates are defined in a single step
— The polysilicon gate can be doped (n+) while is being deposited to lower its parasitic resistance (important in high speed fine line processes)

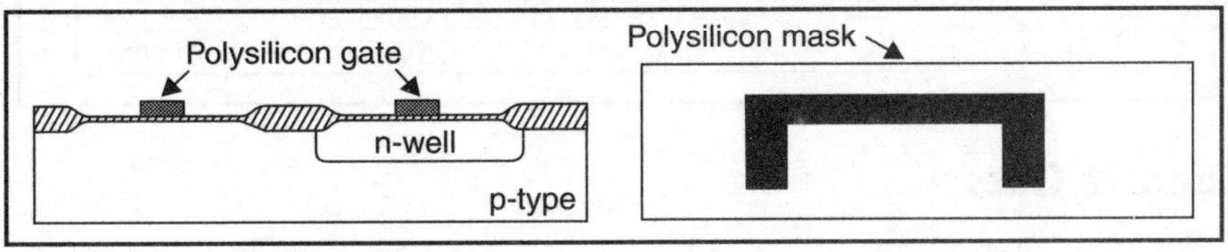

2.1.8 PMOS Formation

— Photoresist is patterned to cover all but the p^+ regions
— A boron ion beam creates the p^+ source and drain regions
— The polysilicon serves as a mask to the underlying channel
- This is called a self-aligned process
- It allows precise placement of the source and drain regions

— During this process the gate gets doped with p-type impurities
- Since the gate had been doped n-type during deposition, the final type (n or p) will depend on which dopant is dominant

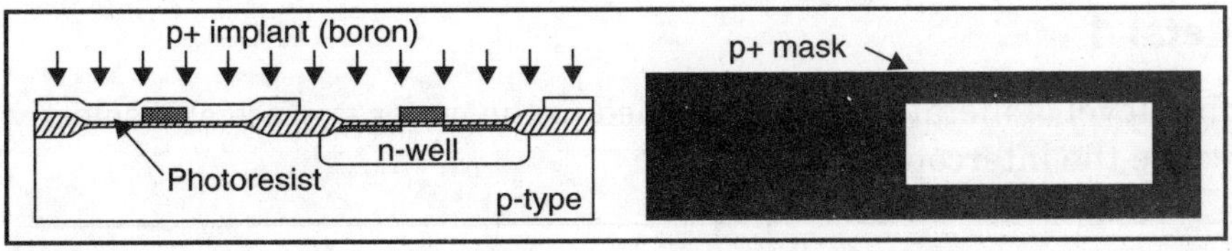

2.1.9 NMOS Formation

— Photoresist is patterned to define the n^+ regions
— Donors (arsenic or phosphorous) are ion-implanted to dope the n^+ source and drain regions
— The process is self-aligned
— The gate is n-type doped

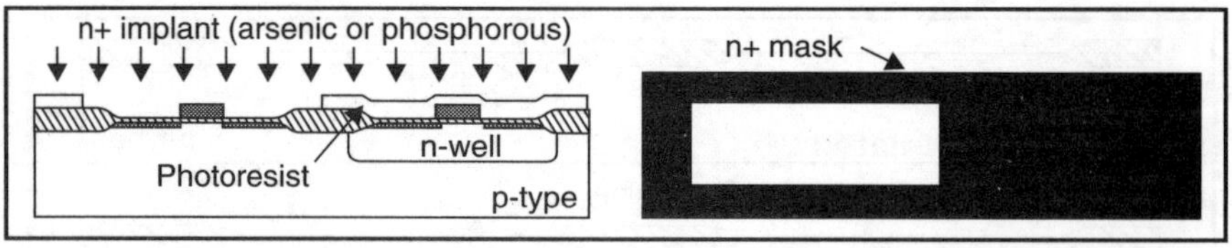

2.1.10 Annealing

- After the implants are completed a thermal annealing cycle is executed
- This allows the impurities to diffuse further into the bulk
- After thermal annealing, it is important to keep the remaining process steps at as low temperature as possible

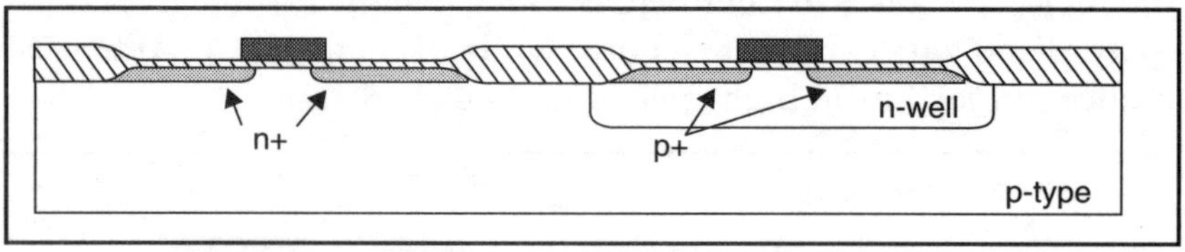

2.1.11 Contact Cuts

- The surface of the IC is covered by a layer of CVD oxide
  - The oxide is deposited at low temperature (LTO) to avoid that underlying doped regions will undergo diffusive spreading
- Contact cuts are defined by etching SiO_2 down to the surface to be contacted
- These allow metal to contact diffusion and/or polysilicon regions

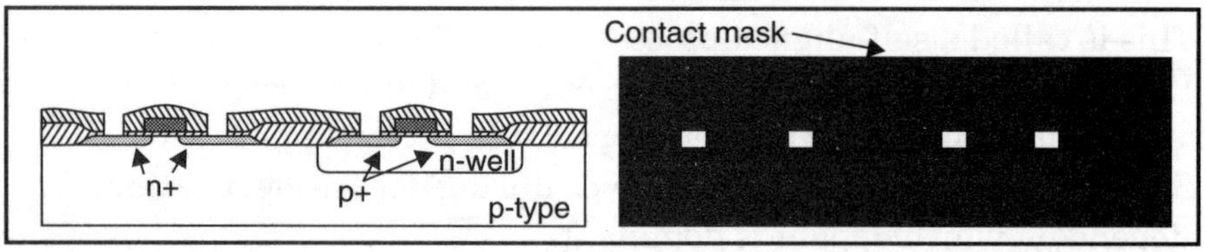

2.1.12 Metal 1

- A first level of metallization is applied to the wafer surface and selectively etched to produce the interconnects

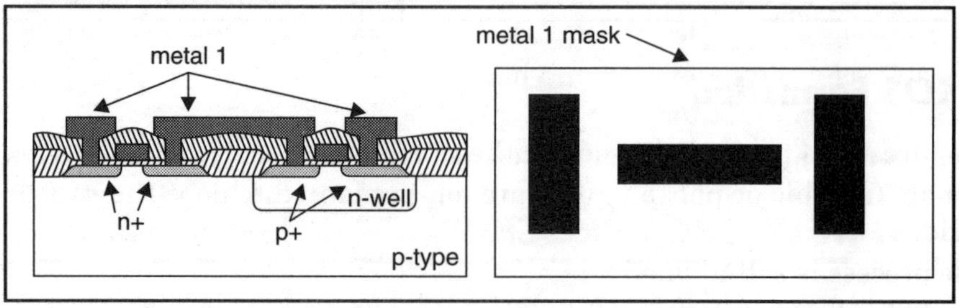

2.1.13 Metal 2

— Another layer of LTO CVD oxide is added
— Via openings are created
— Metal 2 is deposited and patterned

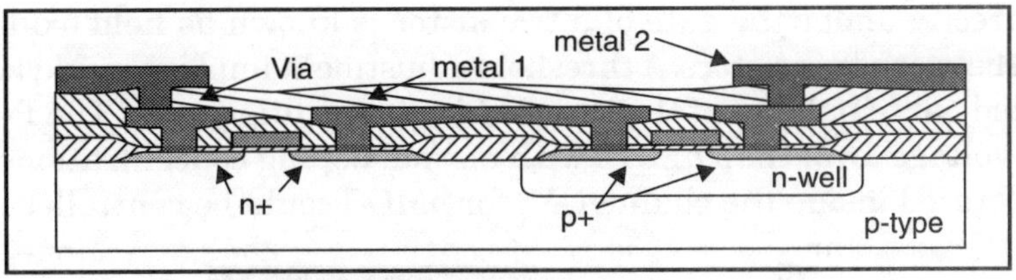

2.1.14 Over Glass and Pad Openings

— A protective layer is added over the surface
— The protective layer consists of
  • A layer of SiO_2
  • Followed by a layer of silicon nitride
— The SiN layer acts as a diffusion barrier against contaminants (passivation)
— Finally, contact cuts are etched, over metal 2, on the passivation to allow for wire bonding.

Advanced CMOS processes:
• Shallow trench isolation
• n^+ and p^+-doped polysilicon gates (low threshold)
• source-drain extensions LDD (hot-electron effects)
• Self-aligned silicide (spacers)
• Non-uniform channel doping (short-channel effects)

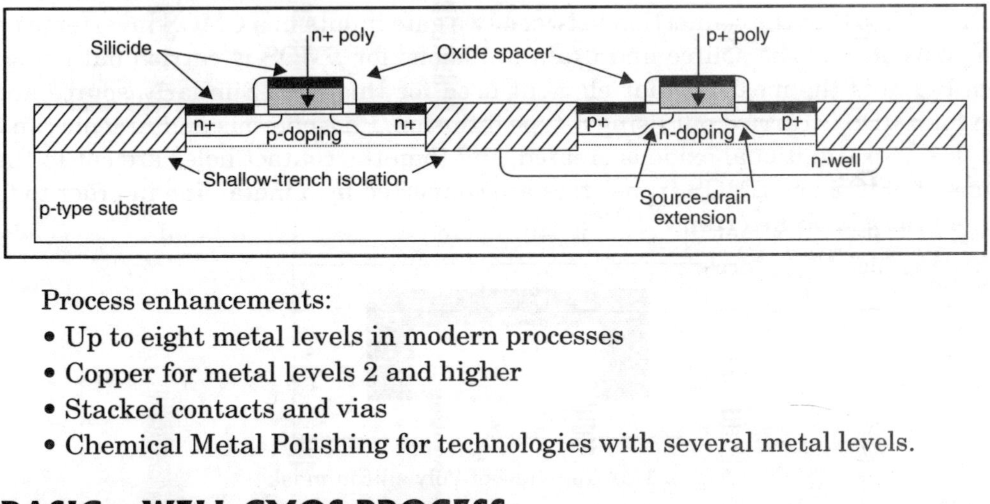

Process enhancements:
• Up to eight metal levels in modern processes
• Copper for metal levels 2 and higher
• Stacked contacts and vias
• Chemical Metal Polishing for technologies with several metal levels.

2.2 BASIC *n*-WELL CMOS PROCESS

In a standard *n*-well process, one of the first things made is the *n*-well in a *p*-type substrate. Once the *n*-well is created, the active areas can be defined. The MOSFET is build within this active area. A very thin layer of silicon dioxide is grown on the surface. This will be used to insulate the gate from the surface. The thin layer of SiO_2 is grown and covered with Si_3N_4. This will act as a mask during the subsequent channel stop implant and field oxide growth. The channel stop implant is to prevent conduction between unrelated transistor source/drains. A thick additional layer oxide grows in both directions vertically where Si_3N_4 is absent. Layer of silicon dioxide under the polysilicon gate (which will be created later) is known as gate oxide and that is not directly under the gate of a transistor is known as field oxide. The field oxide provides isolation between transistors. A threshold adjustment implant would be the next process step. This is carried out to balance of the threshold voltage differences. The P-MOS results in a higher threshold voltage level than *n*MOS with normal doping concentrations. With additional negative charges buried inside the channel, V_T for *p*MOS could be controlled.

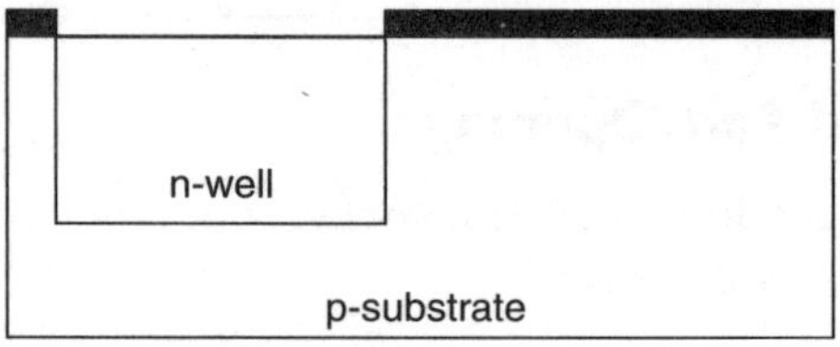

Fig. 2.1(*a*). Formation of *n*-well

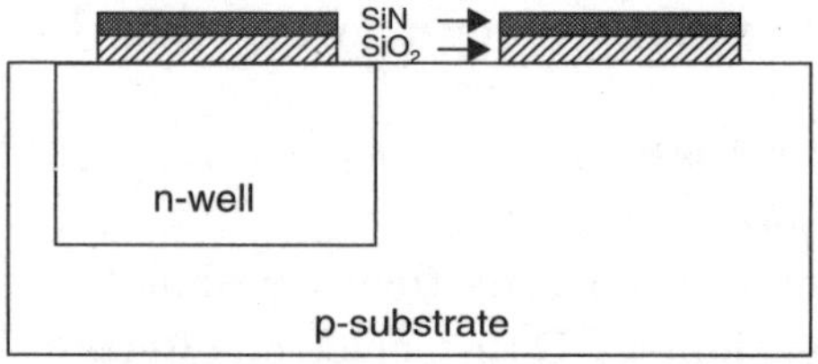

Fig. 2.1(*b*). Gate oxide covered with silicon nitride in the active areas

Polysilicon deposition is carried out and gate definition is then completed using the mask in Fig. 2.1(*c*). Note that the connection between two gate inputs in a CMOS inverter is achieved using the polysilicon. The source and drain diffusions for *p*MOS is carried out using *p*-type diffusion. Boron is the most popular element used for this step. Similarly, source and drain diffusion for *n*MOS is carried out using *n*-type diffusion. Phosphorous and Arsenic can both be used for this step. Additional oxide is created, and then the contact holes are cut in the holes. The drains of *p*MOS and *n*MOS transistors are connected by a metal line in order to take the output from the CMOS inverter.

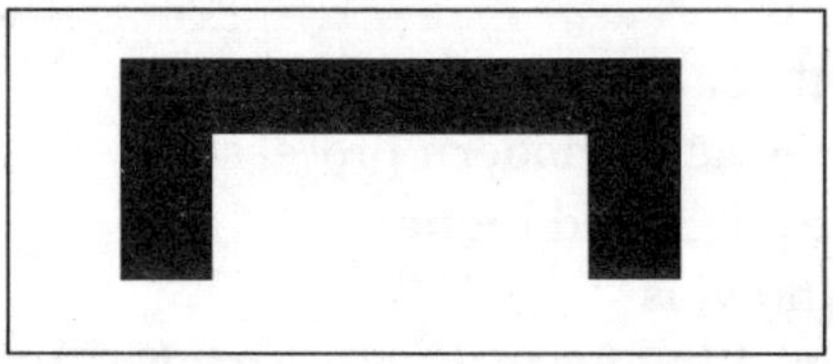

Fig. 2.1(*c*). Top view of Poly silicon mask

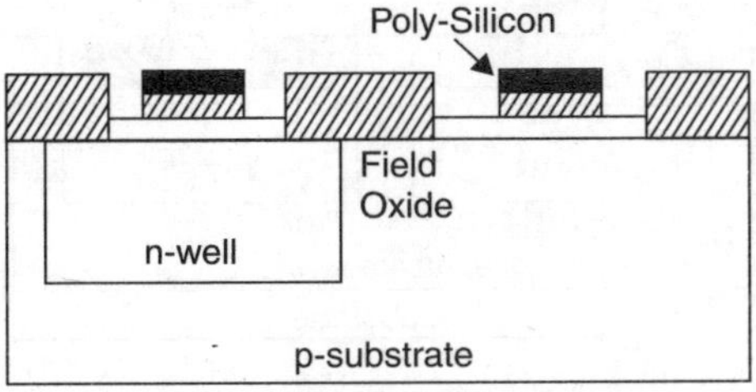

Fig. 2.1(*d*). Poly silicon gate definition is completed

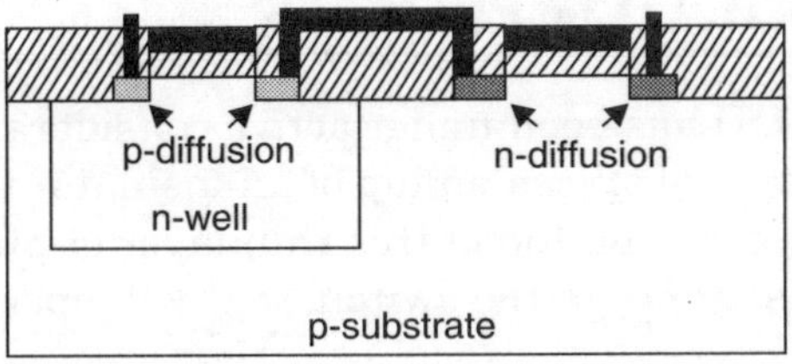

Fig. 2.1(*e*). Transistor source/drain diffusion is completed

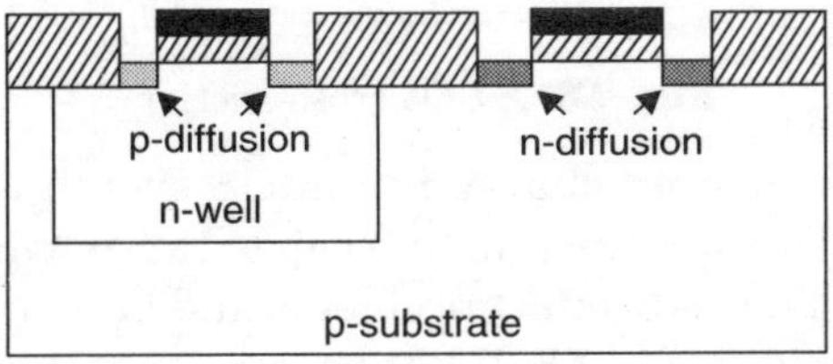

Fig. 2.1(*f*). Cross section of a CMOS inverter in an *n*-well process

2.3 *p*-WELL PROCESS

Prior to the *n*-well process *p*-well process was popular. P-well process is preferred in circumstances where balanced characteristics of the *n*MOS and *p*MOS are needed. It has been observed that the transistors in the native substrate tend to have better characteristics than that was made in a well. Because *p* devices inherently have lower gain than devices, *n*-well process amplifies this difference while a *p*-well process moderates the difference. The standard *p*-well process steps are is similar to *n*-well process, except that a *p*-well is implanted instead of an *n*-well as a first step. Once the *p*-well is created, the active areas and subsequently poly gates can be defined. Later diffusions can be carried out to create source and drain regions. Finally, metal is deposited and patterned for contacts.

2.4 TWIN TUB PROCESS

It is also possible to create both a *p*-well and as *n*-well for the *n*-MOSEFT's and *p*-MOSFET respectively in the twin well or twin tub technology. Such a choice means that the process is independent of the dopant type of the starting substrate (provided it is only lightly doped).

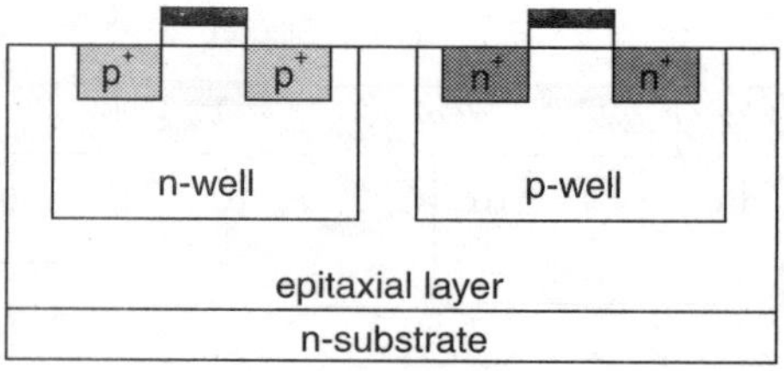

Fig. 2.2. A simplified sketch of twin-well CMOS process cross-section

2.5 SILICON-ON-INSULATOR PROCESS

Silicon-on-Insulator (SOI) has been under active consideration for the last many years. SOI refers to placing a thin layer of silicon on top of an insulator such as silicon oxide or glass. The transistors would then be built on top of this thin layer of SOI. The basic idea is that the SOI layer will reduce the capacitance of the switch, so it will operate faster.

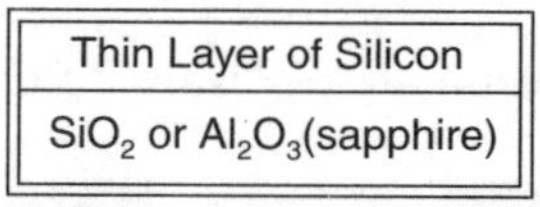

Fig. 2.3. A SOI cross-section

One of the areas that can store charge in a MOS switch is the silicon substrate itself. Since an insulator act as a substrate, the junction capacitance will be eliminated and the MOS transistor will operate faster. The body effect problems and latch up problems are eliminated in SOI. One of the first early applications of SOI has been in memories for space application; since the memories built on SOI were perceived to be more resistance to soft error rate. Soft error rate refers to upset of data in the memory by cosmic rays and background radioactive material. Due to the absence of wells, a closer transistor packing can be achieved in SOI. The major disadvantages of SOI are that it presents lower gains and higher cost. Protecting diodes at inputs cannot be realized in SOI due to absence of silicon substrates.

2.6 LAYOUT DESIGN RULES

Layout design rules describe how small features can be and how closely they can be packed in a particular manufacturing process.

Mead and Conway popularized λ-based design rules based on a single parameter, λ, which characterizes the resolution of the process.

λ is generally half of the minimum drawn transistor channel length. This length is the distance between the source and drain of a transistor and is set by the minimum width of a polysilicon wire.

λ based design rules have become slightly more conservative for more advanced submicron processes, because they round dimensions up to an integer multiple of λ.

Designers often describe a process by its feature size. Feature size refers to minimum transistor length, so λ is half of the feature size.

Simplifed λ-based design rules:

• A conservative but easy to we set of design rules for layouts with two metal layers in an *n*-well process is as follows:

(*i*) Metal and diffusion have minimum width and spacing of 4λ.

(*ii*) Contacts are 2λ × 2λ and must be surrounded by 1 λ on the layers above and below.

(*iii*) Polysilicon uses a width of 2λ.

(*iv*) Polysilicon overlaps diffusion by 2λ where a transistor is desired and has a spacing of 1λ away where no transistor is desired.

(*v*) Polysilicon and contacts have a spacing of 3λ from other polysilicon or contacts.

(*vi*) N-well surrounds *p*MOS transistors of 6λ and avoids *n*MOS transistors of 6λ.

(*vii*) *p*MOS transistors are often wider than *n*MOS transistors because holes move more slowly than electrons so the transistor has to be wider to deliver the same current.

λ-based MOSIS CMOS Design Rules —

• The MOSIS service is a low cost prototyping service that collects designs from academic, commercial and government customers. MOSIS has developed a set of scalable λ-based design rules that cover a wide range of manufacturing processes. There rules are divided by two sets of scalable design rules (*i*) process front end which is the front-end of the line (as in production line) or FEOL and the rules for the process back-end (BEOL), metal interconnect and rules for the pad layer and glass layer.

• **MOSIS Scalable CMOS rules — Process Front end:**

Layer	**Rule**	**Explanation**	**Value/λ**
Well (CWN, CWP)	1.1	minimum width	10
	1.2	minimum space (different potential)	9
	1.3	minimum space (same potential)	0 or 6
	1.4	minimum space (different well type)	0
Active (CAA)	2.1/2.2	minimum width/space	3
	2.3	source/drain active to well edge space	5
	2.4	substrate/well contact active to well edge space	3
	2.5	minimum space between active layers (different implant type)	0 or 4
Poly (CPG)	3.1/3.2	minimum width/space	2
	3.3	minimum gate extension of active layer	2
	3.4	minimum active extension of poly	3
	3.5	minimum field poly to active space	1
Select (CSN, CSP)	4.1	minimum select spacing to channel of transistor[1]	3
	4.2	minimum select overlap of active	2
	4.3	minimum select overlap of contact	1
	4.4	minimum select width and spacing[2]	2

Poly contact (CCP)	5.1(*a*)	exact contact size	2 × 2
	5.2(*a*)	minimum poly overlap	1.5
	5.3(*a*)	minimum contact spacing	2
Active contact (CCA)	6.1(*a*)	exact contact size	2 × 2
	6.2(*a*)	minimum active overlap	1.5
	6.3(*a*)	minimum contact spacing	2
	6.4(*a*)	minimum space to gate of transistor	2

1. To ensure source and drain width.
2. Different select types may touch but not overlap.

- **MOSIS Scalable CMOS rules — Process back end:**

Layer	**Rule**	**Explanation**	**Value/λ**
Metal 1 (CMF)	7.1	minimum width	3
	7.2(*a*)	minimum space	3
	7.2(*b*)	minimum space (for minimum-width wires only)	2
	7.3	minimum overlap of poly contact	1
	7.4	minimum overlap of active contact	1
Via 1 (CVA)	8.1	exact size	2 × 2
	8.2	minimum via spacing	3
	8.3	minimum overlap by metal 1	1
	8.4	minimum spacing to contact	2
	8.5	minimum spacing to poly or active edge	2
Metal 2 (CMS)	9.1	minimum width	3
	9.2(*a*)	minimum space	4
	9.2(*b*)	minimum space (for minimum-width wires only)	3
	9.3	minimum overlap of via 1	1
Via 2 (CVS)	14.1	exact size	2 × 2
	14.2	minimum spacing	3
	14.3	minimum overlap by metal 2	1
	14.4	minimum spacing to via 1	2
Metal 3 (CMT)	15.1	minimum width	6
	15.2	minimum space	4
	15.3	minimum overlap of via 2	2

- **MOSIS Scalable CMOS rules — The pads and overglass (panivation):**

Layer	**Rule**	**Explanation**	**Value/λ**
Glass (COG)	10.1	minimum bonding — pad width	100 μm × 100 μm
	10.2	minimum probe — pad width	75 μm × 75 μm

10.3	pad overlap of glass opening	6 μm
10.4	minimum pad spacing to unrelated metal 2 (or metal 3)	30 μm
10.5	minimum pad spacing to unrelated metal 1, poly, or active	15 μm

Steps : CMOS *n*-well processes Transistor

1. Well

- The *n*-well is usually a deeper implant (especially a deep *n*-well) than the transistor source/drain implants, and therefore, it is necessary for the outside dimension to provide sufficient clearance between the *n*-well edges and the adjacent n^+ diffusions. The inside clearance is determined by the transition of the field oxide across the well boundary.
- The *n*-well sheet resistance can be several KW per square, it is necessary to thoroughly ground the well. This will prevent exercise voltage drops due to well currents.
- A conservative guideline to place a top adjacent to every source connected to V_{DD} or GND. The guidelines are:
  (*i*) Every well should have at least one tap.
  (*ii*) All substrate and well taps should connect direclty to the appropriate supply in metal.
  (*iii*) A tap should be placed for every 5–10 transistor or every 25–100 μm (this distance is process dependent).
  (*iv*) *n*MOS transistors should be clustered together near *a*ND and *p*MOS transistors should be clustered together near V_{DD}, avoiding convoluted structures that intertwine *n*MOS and *p*MOS transistors checker board pattern.
- Where wells are connected to different potentials (in analog circuits), the spacing rules may differ from equipotential wells (in digital logic circuits).

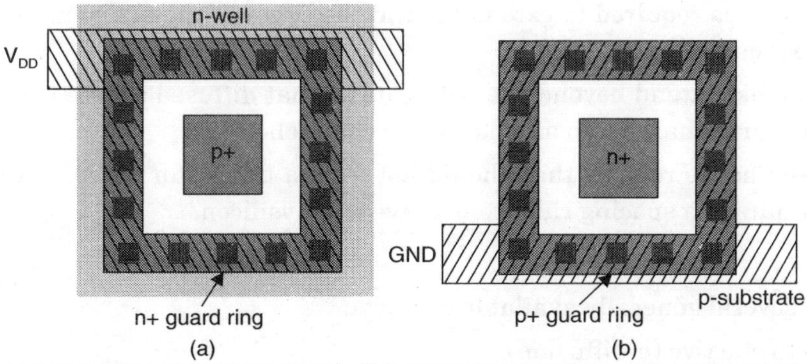

Guard rings

2. Transistor

CMOS transistors are generally defined by at least four physical mark.

(*a*) **Active :**

- They are also called diffusion, different or thinox.
- The active mark defines all areas where either *n*- or *p*-type diffusion is to be placed, or where the gates of transistors are to be placed.
- The polysilicon gates of transistors are defined by the logical AND of the poly Si mark and the active mark, *i.e.,* where poly Si crosses diffusion.

(*b*) **Select Layer :**

- *n*-select (also called *n*-implant, nimp, or *n*-plus).
- *p*-select (also called *p*-implant, pimp or *p*-plus).
- The select layers define what type of diffusion is required.
- *n*-select surrounds active regions where *n*-type diffusion is required, *p* select surrounds areas where *p*-type diffusion is required.
- *n*-diffusion areas inside *p*-well regions define *n*-MOS transistors.
- *n*-diffusion areas inside *n*-well regions define *n*-well contacts.
- *p*-diffusion areas inside *n*-well regions define *n*-MOS transistors.
- *p*-diffusion areas inside *p*-well regions define *p*-well contacts.
- Frequently, design systems will only define *n*-diffusion and *p*-diffusion to reduce the complexity of the process.
- The appropriate selects are generated automatically.
- That is, *n* diffusion will be converted automatically into active with an overlapping rectangle or polygon of *n*-select.

(*c*) **Poly :**

- It is evential for the poly to be completely cross active ; otherwise the transistor that has been created will be shorted by a diffusion path between source and drain.
- Hence poly is required to extend beyond the edges of the active area. This is called gate extension.
- Active must extend beyond the poly gate so that diffused source and drain regions exist to carry charge into and out of the channel.
- Poly and active regions that should not form a transistor must be kept separated, this results in a spacing rule from active to polysilicon.

(*d*) **Contact :**

There are several generally available contacts:

- Metal to *p*-active (*p*-diffusion).
- Metal to *n*-active (*n*-diffusion).
- Metal to polysilicon.
- Metal to well or substrate.

Depending on the process, other contacts such as burried polysilicon – active contacts may be allowed for local interconnect.

Each isolated well must be tied to the appropriate supply voltage, because the substrate is divided into well regions.

So, *n*-well must be tied to V_{DD} and the substrate or *p*-well must be tied to GND with well or substrate contacts. A split or merged contact is equivalent to two adjacent contacts to *n*-active and *p*-active strapped together with metal. This structure is used to tie transistor sources to the substrate or *n*-well and simultaneously to GND or V_{DD} as shown at the source of the *n* transistor in Fig. below.

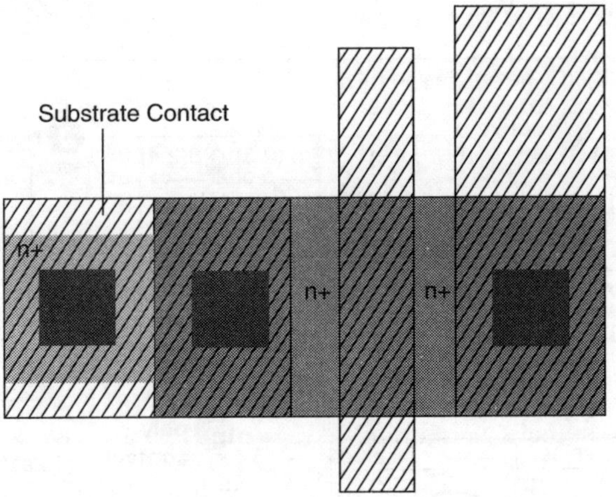

Substrate contact

(*e*) **Metal Rules:**

Metal spacing may vary with the width of the metal line (so called fat metal rules). That is, above some metal wire width, the minimum spacing may be increased. This is due to each characteristics of small versus large metal wires.

Maximum Metal Width Rule:

Single metal wires cannot be greater than a certain width. If wider wires are desired, they are constructed by paralleling a number of smaller wires and adding checkerboard links to tie the wires together.

(*f*) **Via Rules:**

Processes may vary in whether they allow stacked vias to be placed over polysilicon and diffusion regions. Some processes allow vias to be placed within there areas, but do not allow the vias to straddle the boundary of polysilicon or diffusion.

Most modern plannarized processes allow for stacked vias whick reduces the area required to pass from a lower level metal to a high-level metal.

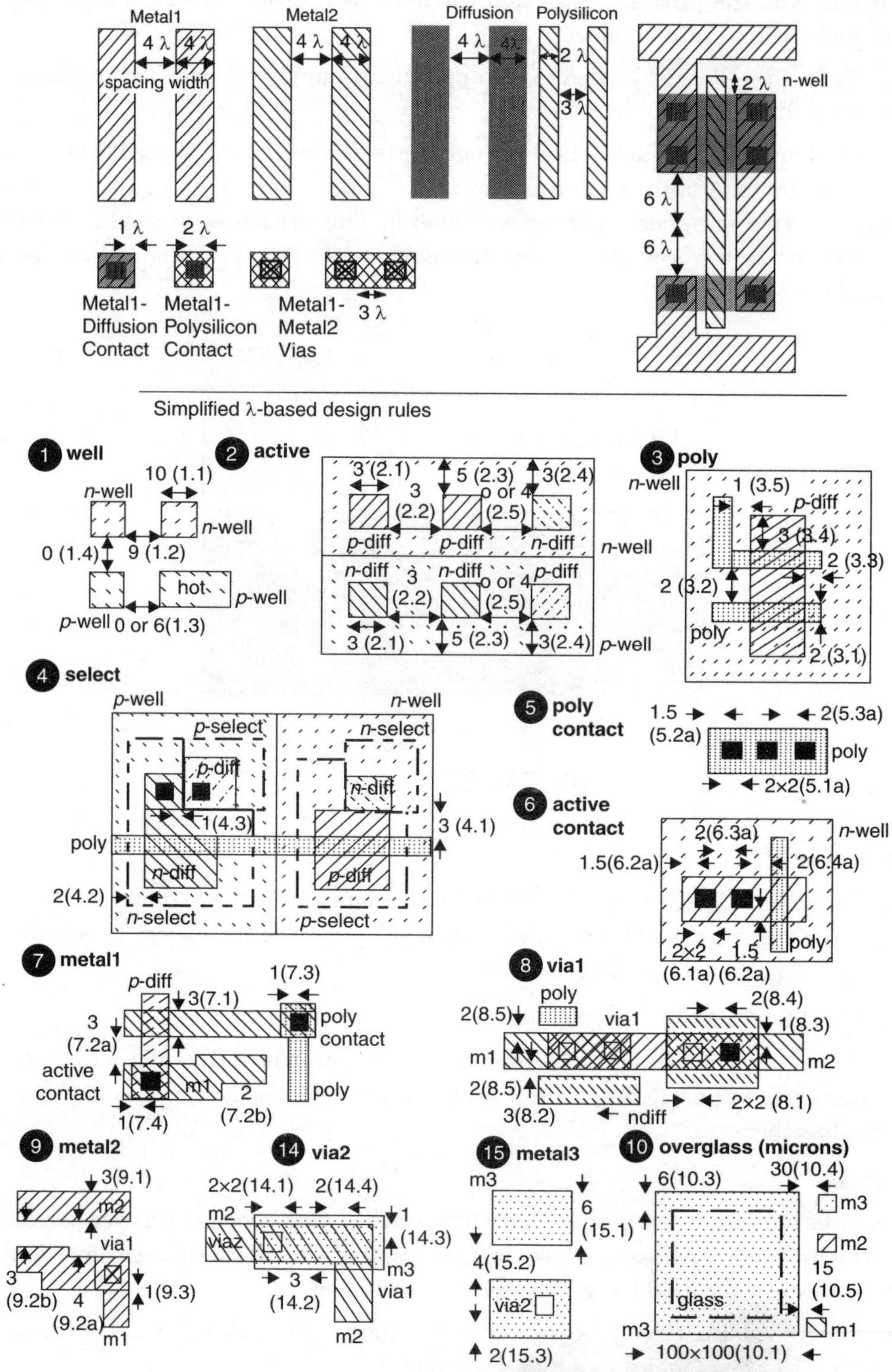

Fig. 2.4. The MOSIS scalable CMOS design rules (rev. 7). Dimensions are in λ. Rule numbers are in parentheses (missing rule sets 11–13 are extensions to this basic process).

2.7 CMOS INVERTER DESIGNS

CMOS inverter designs are based on the following entities:

Substrate and/or *wells*, being *p*-type (for NMOS devices) and *n*-type (for PMOS)

- *Diffusion regions* (n^+ and p^+) defining the areas where transistors can be formed.

  These regions are often called the *active areas.*

  Diffusion of an inverse type are needed to implement contacts to the wells or to the substrate. These are called *select regions.*
- One or more *polysilicon* layers, which are used to form the gate electrodes of the transistors (but serve as interconnect layers as well).
- A number of *metal interconnect* layers.
- *Contact and via* layers to provide interlayer connections.

  A layout consists of a combination of polygons, each of which is attached to a certain layer. The functionality of the circuit is determined by the choice of the layers, as well as the interplay between objects on different layers. For instance, an MOS transistor is formed by the cross-section of the diffusion layer and the polysilicon layer. An interconnection between two metal layers is formed by a cross-section between the two metal layers and an additional contact layer.

1. *Transistor Rules* : A transistor is formed by the overlap of the active and the polysilicon layers. From the intralayer design rules, it is already clear that the minimum length of a transistor equals. 0.24 mm (the minimum width of polysilicon), while its width is at least 0.3 mm (the minimum width of diffusion). Extra rules include the spacing between the active area and the well boundary, the gate overlap of the active area, and the active overlap of the gate.
2. *Contact and Via Rules* : A contact (which forms an interconnection between metal and active or polysilicon or a via (which connects two metal layers) is formed by overlapping the two interconnecting layers and providing a contact hole, filled with metal, between the two. In our process, the minimum size of the contact hole is 0.3 mm, while the polysilicon and diffusion layers have to extend at least over 0.14 mm beyond the area of the contact hole. This sets the minimum area of a contact to 0.44 mm and 0.44 mm. This is larger than the dimensions of a minimum-size transistor! Excessive changes between interconnect layers in routing are thus to be avoided. The figure, furthermore, points out the minimum spacings between contact and via holes, as well as their relationship with the surrounding layers.
3. *Well and Substrate Contacts* : For robust digital circuit design, it is important for the well and subtrate regions to be adequately connected to the supply voltages.

   Failing to do so results in a resistive path between the substrate contact of the transistors and the supply rails, and can lead to possibly devastating parasitic effects, such as latchup. It is therefore advisable to provide numerous substrate (well) contacts spread over the complete region. To establish an ohmic contact between a supply rail, implemented in metal 1, and a *p*-type material, a p^+ diffusion region must be provided. This is enabled by the *select* layer, which reverses the type of diffusion. A number of rules regarding the use of the *select layer.*

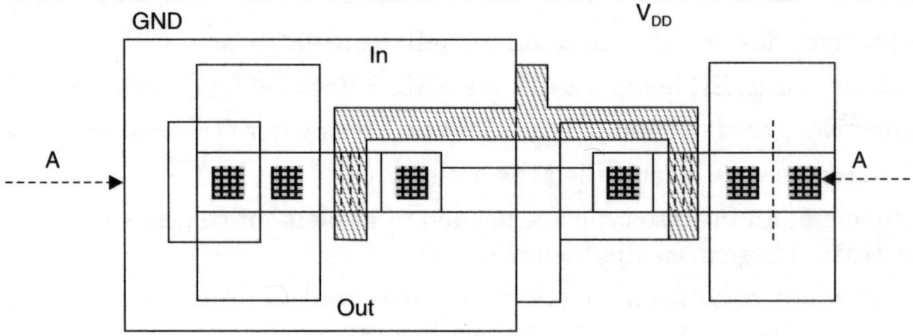

(a) Layout

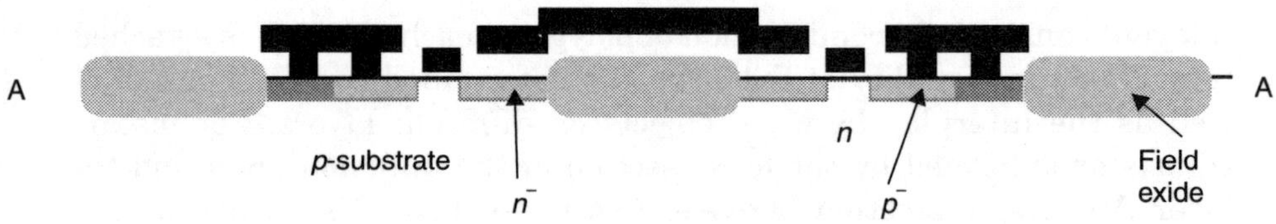

(b) Cross-section along A-A

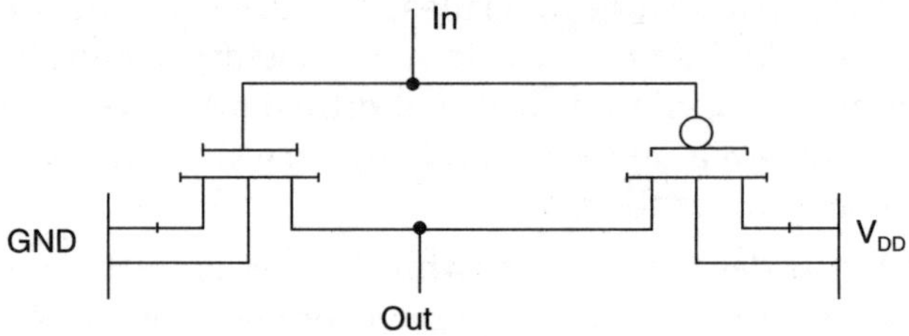

(c) Circuit diagram

Fig. 2.5. A detailed layout example, including vertical process cross-section and circuit diagram

Chapter 3

INVERTER

3.1 INTRODUCTION

The inverter is universally accepted as the most basic logic gate doing a Boolean operation on a single input variable. Figure 3.1 depicts the symbol, truth table and a general structure of a CMOS inverter. As shown, the simple structure consists of a combination of an pMOS transistor at the top and a nMOS transistor at the bottom.

Truth Table :

IN	OUT
'0'	'1'
'1'	'0'

Fig. 3.1

Block Diagram :

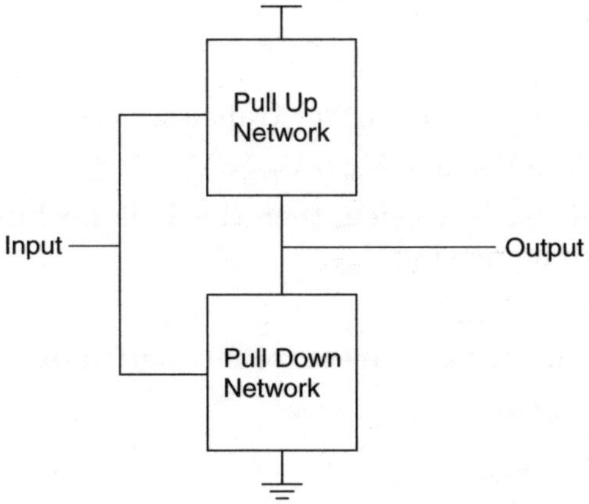

Fig. 3.2

The truth table of the inverter shows a logic '1' output corresponding to a logic '0' in the input. This can be ensured by the p-transistor whose source is connected to a logic '1' source (V_{DD}) and gate is provided a logic '0' stimulus. As in this condition, the n-transistor is OFF (that is, there is no current path between the output node and ground node) and the p-transistor channel is in the linear or resistive region of operation, one obtains perfect logic '1' at the output node. Similarly, a logic '0' output will result from a logic '1' input. The *n*MOS transistor connected in the bottom realizes this when its gate is given a logic '1' input and its source is connected to logic '0' or ground (V_{SS}). In this case, the *n*MOS transistor channel acts like a wire resulting in logic '0' at the output while the *p*MOS transistor channel is simply devoid of any conductive channel.

The inverter can best be considered as the central part of digital designs. A thorough understanding of its operation and properties is required to design more complex structures like NAND and NOR gates, adders and multipliers.

3.2 RESISTIVE LOAD INVERTER

The MOSFET is used in enhancement type *n*MOS and the input is applied to the gate of the *n*MOS transistor which is also known as driver.

The load is a simple linear resistor R which is connected between the drain and the supply voltage V_{DD}. The substrate of the driver is grounded and the source is also kept at a ground potential. Since $V_{SB} = 0$, the threshold voltage of the driver is V_{TO} (V_T at $V_{SB} = 0$).

So, $I_L = I_D$ in steady state DC operation, where channel length modulation is also neglected.

$$\therefore \quad I_L = \frac{V_{DD} - V_{out}}{R_L}$$

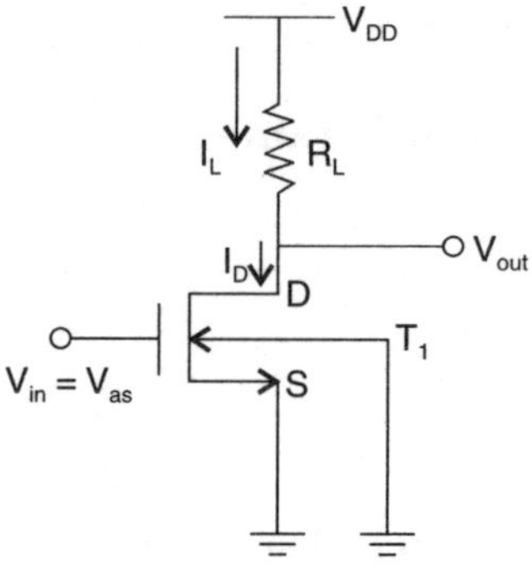

Fig. 3.3

$$\text{Now, } I_D = \frac{K_n}{2}\,[2\,(V_{in} - V_{TO})\,V_{out} - V_{out}^2]$$

When $V_{in} - V_{TO} > V_{out}$ (linear region)

$$= \frac{K_n}{2}\,[V_{in} - V_{TO}]^2$$

When $V_{in} - V_{TO} < V_{out}$ (saturation region).

$$\text{Where } K_n = \frac{\mu_n\, C_{ox}\, W}{L}$$

Case 1. When $V_{as} = V_{in} < V_{TO}$, the driver transistor is in cut off.

Case 2. When $V_{as} = V_{in} > V_{TO}$ and $V_{in} - V_{TO} < V_{as}$.

When a small V_{DS} is applied, then the T_1 is working in linear region where I_D increases with increasing V_{DS}.

Case 3. $V_{DS} = V_{out} > V_{in} - V_{TO}$

The driver saturates and the current equation in saturation region.

Transfer characteristics curve:

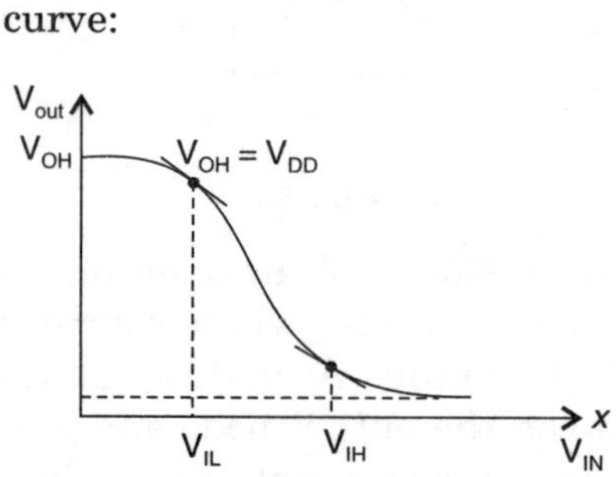

Fig. 3.4

Calculation of V_{OH}:

When $V_{in} < V_{TO}$, then driver transistor is cut off.

So, $I_D = 0, V_{out} = V_{OH}$.

$$I_L = I_D \; ; \; \frac{V_{DD} - V_{out}}{R_L} = I_D = 0.$$

When $I_D = 0$, then $V_{out} = V_{OH}$.

$$\text{Now, } \frac{V_{DD} - V_{OH}}{R_L} = 0, \qquad \boxed{V_{DH} = V_{DD}}$$

Calculation of V_{OL}:

Assuming that the input voltage is high $V_{in} = V_{OH} = V_{DD}$ and $V_{out} = V_{OL}$.

Since $V_{out} = V_{OL}$ and $V_{in} - V_{TO} > V_{out}$, the transistor operates in linear region.

$$\text{Then, } \quad \frac{V_{DD} - V_{out}}{R_L} = \frac{K_n}{2} [2(V_{in} - V_{TO}) V_{out} - V_{out}^2]$$

Substituting, $V_{DD} = V_{in}$ and $V_{out} = V_{OL}$, we get

$$\frac{V_{DD} - V_{OL}}{R_L} = \frac{K_n}{2} [2(V_{DD} - V_{TO}) V_{OL} - V_{OL}^2]$$

$$\text{or, } \quad V_{OL}^2 - 2\left(V_{DD} - V_{TO} + \frac{1}{K_n R_L}\right) V_{OL} + \frac{1}{K_n R_L} V_{DD} = 0.$$

Since V_{OL} is a positive quantity and lies between 0 and V_{DD}. So, V_{OL} is obtained as,

$$V_{OL} = V_{DD} - V_{TO} + \frac{1}{K_n R_L} \sqrt{\left(V_{DD} - V_{TO} + \frac{1}{K_n R_L}\right)^2 - \frac{2V_{DD}}{K_n R_L}}.$$

Calculation of V_{IL}:

From the transfer function curve, we observe that V_{IL} is the smaller of the two values for which the slope of the voltage transfer curve becomes equal to (– 1). Also the O/P voltage at V_{IL} is approximately equal to V_{OH}. Since $V_{in} - V_{TO} < V_{out}$, then the T_1 operates in saturation region.

$$\text{Now, } \quad \frac{V_{DD} - V_{out}}{R_L} = \frac{K_n}{2} (V_{in} - V_{TO})^2.$$

Differentiating with respect to V_{in}, we get

$$-\frac{1}{R_L} \cdot \frac{dV_{out}}{dV_{in}} + K_n (V_{in} - V_{TO}).$$

Now substituting $\dfrac{dV_{out}}{dV_{in}} = -1$ and $V_{in} - V_{IL}$, we have,

$$\boxed{V_{IL} = V_{TO} + \frac{1}{K_n R_L}}$$

Now, $$\frac{V_{DD} - V_{out}}{R_L} = \frac{K_n}{2}(V_{IL} - V_{TO})^2$$

or, $$\frac{V_{DD} - V_{out}}{R_L} = \frac{K_n}{2}\left(V_{TO} + \frac{1}{K_n R_L} - V_{TO}\right)^2$$

or, $$\frac{V_{DD} - V_{out}}{R_L} = \frac{K_n}{2} \cdot \frac{1}{K_n^2 R_L^2}$$

or, $$\boxed{V_{out} = V_{DD} - \frac{1}{2K_n R_L}}$$

Calculation of V_{IH}:

We observe that V_{IH} is the larger of the two values of input voltages for which the slope of the voltage transistor curve is – 1.

Also, the O/P voltage at V_{IH} is slightly greater than V_{OL}. Since $V_{in} - V_{TO} > V_{out}$ the transistor is in linear region.

So, $$\frac{V_{DD} - V_{out}}{R_L} = \frac{K_n}{2}[2(V_{in} - V_{TO})V_{out} - V_{out}^2]$$

Differentiating the equation w.r.t. V_{in}, we get :

$$-\frac{1}{R_L} \cdot \frac{dV_{out}}{dV_{in}} = \frac{V_n}{2}\left[2(V_{in} - V_{TO})\frac{dV_{out}}{dV_{in}} - V_{out} - V_{out}\frac{dV_{out}}{dV_{in}}\right]$$

Substituting, $\frac{dV_{out}}{dV_{in}} = 1$ and $V_{in} = V_{IH}$, we get

$$\frac{1}{R_L} = \frac{K_n}{2}[-2(V_{IH} - V_{TO}) + 4V_{out}]$$

or, $$V_{IH} = V_{TO} + 2V_{out} - \frac{1}{K_n R_L}$$

$\therefore$ $$\frac{V_{DD} - V_{out}}{R_L} = \frac{K_n}{2}\left[2\left(V_{TO} + 2V_{out} - \frac{1}{K_n R_L} - V_{TO}\right)V_{out} - V_{out}^2\right]$$

or, $$V_{out} = \sqrt{\frac{2}{3}\frac{V_{DD}}{K_n R_L}}$$

$$\therefore \quad \boxed{V_{IH} = V_{TO} + 2\sqrt{\frac{2}{3}\frac{V_{DD}}{K_n R_L}} - \frac{1}{K_n R_L}}$$

Noise margin, propagation delay and fall time calculation:

At time t_1, the O/P has dropped by 10% of the logic saving ΔV and time t_4 is the time at which the O/P has dropped by 90% of ΔV.

So, the fall time, $t_f = t_4 - t_1$.

At t_3 the O/P is at the 50% point, that is

$$V_{50\%} = \frac{V_{OH} + V_{OL}}{2}$$

$\therefore$ The propagation delay,

$$\tau_{pHL} = t_3.$$

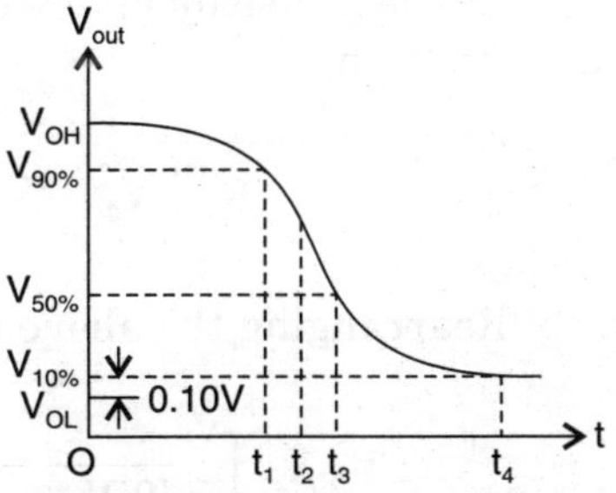

Fig. 3.5. Variation of O/P voltage with time for an inverter

At time t_2, $V_O = V_{DD} - V_{th}$ and for there voltage levels, the transistor changes from saturation region operation to linear region operation.

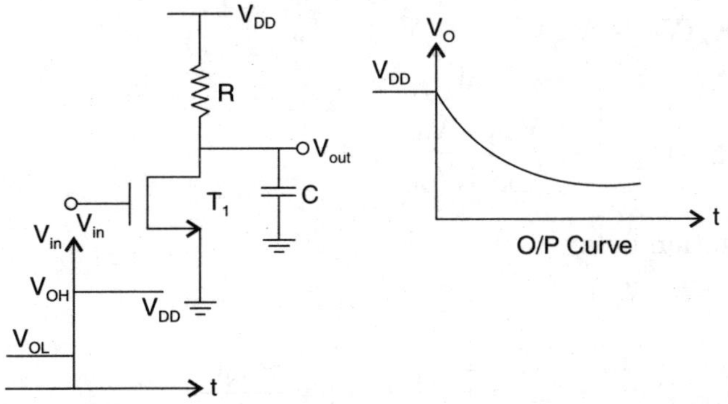

Fig. 3.6

From the above O/P curve, the capacitor current is described by,

$$\frac{K_n}{2}(V_{gs} - V_{th})^2 = -C\frac{dVe}{dt}$$

The drain current is constant and the capacitor discharges at a constant rate until the voltage at the capacitor reaches 4 V (if $V_{DD} = 5$ V).

Then $V_{D3} \leq V_{as} - V_{th}$ and T_1 enters the linear region.

So, the time t_2 is required for the transistor to reach the linear region is,

$$t_2 = \frac{2CV_{th}}{K_n(V_{DD} - V_{th})^2} = 2\,R_{ON}\,C\,.\,\frac{V_{th}}{(V_{DD} - V_{th})}$$

where, ON resistance, $R_{ON} = \dfrac{1}{K_n(V_{DD} - V_{th})}$

If, $V_C = V_{gs} - V_{th}$ then,

$$t_2 \approx \frac{C}{8\,K_n} = \frac{R_{ON}C}{2}$$

As the transistor enters the linear region of operation the equation describing the discharge takes the form,

$$K_n\left(V_{gs} - V_{th} - \frac{V_C}{2}\right) = C\,\frac{dV_C}{dt} \text{ with } V_{DS} = V_C.$$

Rearranging the above equation with $V_{as} = V_{DD}$ and integrating, we have,

$$\int_{V_2}^{V_3} \frac{dV_C}{[2(V_{DD} - V_{th}) - V_C]\,V_C} = \int_{t_2}^{t_3} \frac{K_n}{2C}\,dt.$$

where, $V_2 = V_C(t_2) = V_{DD} - V_{th}$; $V_3 = V_C(t_3) = 0.5\,(V_{DD} + V_{OC})$.

Solving the integral equation we can get,

$$t_3 - t_2 = \frac{C}{K_n(V_{DD} - V_{th})} \ln\left[4\left(\frac{V_{DD} - V_{th}}{V_{DD} + V_{OL}}\right) - 1\right]$$

$$= R_{ON}\,C \ln\left[4\left(\frac{V_{DD} - V_{th}}{V_{DD} + V_{OL}}\right) - 1\right]$$

∴ The propagation delay,

$$T_{PHL} = t_3 = t_3 - t_2 + t_2$$

$$= R_{ON}\,C \ln\left[4\left(\frac{V_{DD} - V_{th}}{V_{DD} + V_{OL}}\right) - 1\right] + \frac{R_{ON}C}{2}$$

$$= R_{ON}\,C \ln\left[\left\{4\left(\frac{V_{DD} - V_{th}}{V_{DD} + V_{OL}}\right) - 1\right\} + \frac{1}{2}\right]$$

Fall time:

$$t_f = (t_4 - t_1)$$

$$= (t_4 - t_2) + (t_2 - t_1)$$

At time $(t_2 - t_1)$, the MOSFET is in saturation region and the current discharging the capacitor is constant.

Now, $$\Delta t = \frac{C\Delta V}{I}$$

∴ $$t_2 - t_1 = \frac{C\{V_{DD} - 0.1\,\Delta V) - (V_{DD} - V_{th})\}}{\frac{K_n}{2}(V_{DD} - V_{th})^2}$$

$$= 2\,R_{ON}\,C\,\frac{(V_{th} - 0.1\,\Delta V)}{V_{DD} - V_{th}}$$

At time $t_4 - t_2$ the MOSFET is operating in the (linear region). So,

$$\int_{V_2}^{V_4} \frac{dV_C}{[2(V_{DD} - V_{th}) - V_C]V_C} = \int_{t_2}^{t_4} \frac{K_n}{2C}\,dt.$$

where, $V_2 = V_C\,(t_2) = V_{DD} - V_{th}$

$V_4 = V_C\,(t_4) = V_{DD} - 0.9\,\Delta V.$

Solving the integral equation, we get,

$$t_4 - t_2 = R_{ON}\,C\,\ln\left[2\left(\frac{V_{DD} - V_{th}}{V_{DD} - 0.9\,\Delta V}\right) - 1\right]$$

Now,

Fall time:

$$t_f = t_4 - t_1$$

$$= (t_4 - t_2) + (t_2 - t_1)$$

$$= R_{ON}\,C\,\ln\left[2\,I\left(\frac{V_{DD} - V_{th}}{V_{DD} - 0.9\,\Delta V}\right) - 1\right] + 2\left(\frac{V_{th} - 0.1\,\Delta V}{V_{DD} - V_{th}}\right)$$

NOISE margin of a resistive load inverter;

$$NM_L = V_{IL} - V_{OL}$$

$$= \left(V_{TO} + \frac{1}{K_n R_L}\right) - \left\{V_{DD} - V_{TO} + \frac{1}{K_n R_L}\sqrt{\left(V_{DD} - V_{TO} + \frac{1}{K_n R_L}\right)^2 - \frac{2V_{DI}}{K_n R_L}}\right\}$$

$$NM_H = V_{OH} - V_{IH}$$

$$= V_{DD} - \left\{V_{TO} + 2\sqrt{\frac{2}{3}\frac{V_{DD}}{K_n R_L}} - \frac{1}{K_n R_L}\right\}$$

Advantages:

(*i*) Design principle very simple.

(*ii*) Low cost.

Disadvantage:

It is not suitable for MOS primary VLSI circuits because of the large area occupied by the load resistor.

3.3 ACTIVE LOAD INVERTER

To solve the problem of a resistive load inverter we design an active load inverter.

The main advantage of using a MOSFET as the load device is that the silicon area occupied by the transistor is usually smaller than that occupied by a comparable resistive load.

Enhancement — Load MOS Inverter :

Depending on the bias voltage applied to its gate terminal, the load transistor can be operated either in the saturation region or in the linear region.

There are two types of load enhancement type inverters:

(*i*) Saturated enhancement *n*MOS load inverter.

(*ii*) Linear enhancement *n*MOS load inverter.

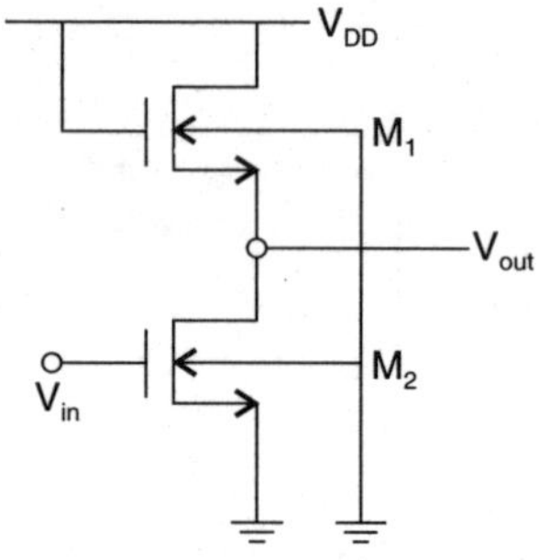

Fig. 3.7. Saturated enhancement type *n*-MOS load inverter

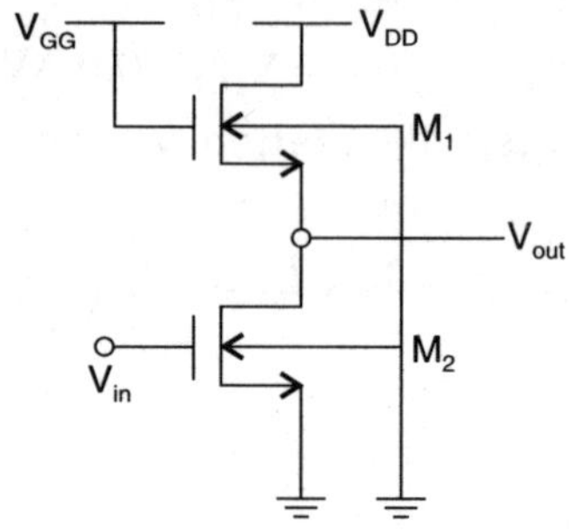

Fig. 3.8. Linear enhancement type *n*-MOS load inverter

The saturated enhancement type load inverter requires a single voltage supply and a relatively simple fabrication process, yet the V_{OH} level is limited to $V_{DD} - V_{T1}$.

But the linear enhancement type load inverter is always biased in the linear region. Thus, the V_{OH} level is equal to V_{DD} and it has higher noise margins compared to saturated enhancement load inverter.

The most significant drawback of this type of configuration is the use of two separate power supply voltage.

But both types of inverter suffer from relatively high stand – by (DC) power dissipation, hence this type of inverter is not used in any large scale digital application.

Depletion–Load *n*MOS Inverter:

The disadvantages of the enhancement type load inverter can be avoided by using a depletion type *n*MOS transistor as the load device.

Advantages :

(*i*) Sharp VTC transition and better noise margins.

(*ii*) Using single power supply.

(*iii*) Smaller overall layout area.

The driver device is an enhancement type *n*MOS transistor with $V_{TO,\,M2} > 0$. Whereas the load is a depletion type *n*MOS transistor, with $V_{TO,\,M1} < 0$.

The current voltage equations to be used for the depletion type load transistor are identical to the enhancement type device with the exception of the negative threshold voltage.

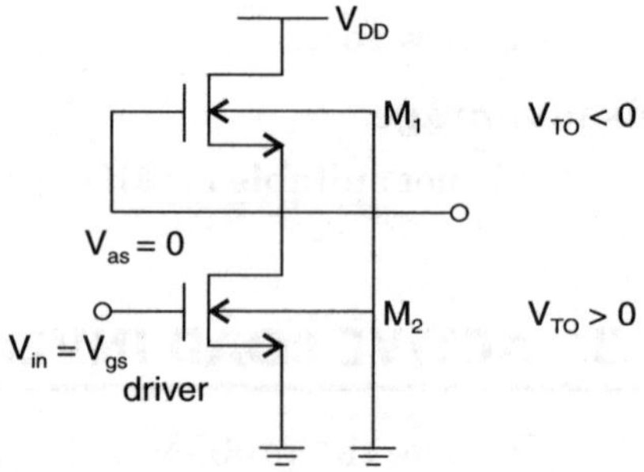

Fig. 3.9

The gate and the source nodes of the load transistor are connected ; hence, $V_{gs}, m_1 = 0$ V always.

$V_{gs, M1} > V_{T, M1}$, because of negative threshold voltage, for this reason, the load device always has a conducting channels regardless of the input/output voltage levels. The operating mode of the load transistor is determined by the output voltage level which is shown in the table given below :

V_{in}	V_{out}	*Driver operating region*	*Load operating region*
V_{OL}	V_{OH}	Cutoff	Linear
V_{IL}	$\approx V_{OH}$	Saturation	Linear
V_{OH}	V_{OL}	Linear	Saturation
V_{IH}	Small	Linear	Saturation

Calculation of V_{OH}:

When $V_{in} < V_{TO\ M2}$ then $M_2 \rightarrow$ cutoff. $I_{D_{M2}} \approx 0$.

But for M_1, M_1 operates in linear region and drain current is also zero. So,

$$I_{D, M_1} = \frac{K_{n1}}{2} [2 \mid V_{th_1} (V_{OH}) \mid . (V_{DD} - V_{OH}) - (V_{DD} - V_{OH})^2] = 0$$

$$\therefore \quad \boxed{V_{OH} = V_{DD}.}$$

Calculation of V_{OL}:

If, $V_{in} = V_{OH} = V_{DD}$, then

$M_1 \rightarrow$ in the saturation region,

$M_2 \rightarrow$ in the linear region,

So, $I_{DM_1} = I_{DM_2}$, then,

$$\frac{K_2}{2} [2 (V_{OH} - V_{TO}) . V_{OL} - V_{OL}^2] = \frac{K_1}{2} [- V_{T_1} (V_{OL})]^2$$

or
$$V_{OL} = V_{OH} - V_{TO} - \sqrt{(V_{OH} - V_{TO})^2 - \left(\frac{K_1}{K_2}\right) \left| V_{T_1} (V_{OL}) \right|^2} .$$

Calculation of V_{IL}:

By definition, the slope of the VTC is equal to (– 1), *i.e.*, $\frac{dV_{out}}{dV_{in}} = -1$ when the input voltage is $V_{in} = V_{IL}$.

$M_1 \rightarrow$ operates in linear region.

$M_2 \rightarrow$ operates in saturation region.

$$\frac{K_{M2}}{2} (V_{in} - V_{TO})^2 = \frac{K_{M1}}{2} [2 \mid V_{T, M_1} (V_{out}) \mid (V_{DD} - V_{out}) - (V_{DD} - V_{out})^2]$$

To satisfy the derivative condition at V_{IL}, we differentiate both sides of the above equation w.r.t. V_{in}

$$K_{M_2}(V_{in}-V_{TO}) = \frac{K_{M_1}}{2}\left[2|V_{T,M_1}(V_{out})|\left(-\frac{dV_{out}}{dV_{in}}\right)+2(V_{DD}-V_{out})\left(-\frac{dV_{T,M_1}}{dV_{in}}\right) - 2(V_{DD}-V_{out})\left(-\frac{dV_{out}}{dV_{in}}\right)\right]$$

$$\frac{dV_{T,M_1}}{dV_{in}} \to 0 \text{ w.r.t. others.}$$

Now, put, $V_{in} = V_{IL}$ and $\frac{dV_{out}}{dV_{in}} = -1$, then,

$$V_{IL} = V_{TO} + \left(\frac{K_{M1}}{K_{M2}}\right)[V_{out} - V_{DD} + |V_{T,load}(V_{out}|].$$

Calculation of V_{IH} :

V_{IH} is the larger of the two voltage points on the VTC at which the slope is equal to (– 1). Since the output voltage corresponding to this operating point, is relatively small. Then,

$M_1 \to$ operates in saturation region,

$M_2 \to$ operates in linear region.

$$\frac{K_{M_2}}{2}[2(V_{in}-V_{TO})V_{out} - V_{out}^2] = \frac{K_{M1}}{2}[-V_{TM1}(V_{out})]^2$$

Differentiating both sides w.r.t. V_{in}, we get,

$$K_{M_2}\left[V_{out} + (V_{in}-V_{to})\left(\frac{dV_{out}}{dV_{in}}\right) - V_{out}\left(\frac{dV_{out}}{dV_{in}}\right)\right]$$

$$= K_{M_1}[-V_{T,M_1}(V_{out})]\left(\frac{dV_{T,M_1}}{dV_{out}}\right)\left(\frac{dV_{out}}{dV_{in}}\right)$$

Now substitute $\frac{dV_{out}}{dV_{in}} = -1$ and $V_{in} = V_{IH}$, then,

$$V_{IH} = V_{TO} + 2V_{out} + \left(\frac{K_{M_1}}{K_{M_2}}\right)[-V_{TM_1}(V_{out})]\left(\frac{dV_{T,M_1}}{dV_{out}}\right)$$

where, $$\frac{dV_{T,M_1}}{dV_{out}} = \frac{r}{2\sqrt{|2\phi_F| + V_{out}}}$$

Propagation Delay, Noise Margin, Rise Time and Fall Time :

$$\tau_{PHL} = \frac{\tau}{K_{M_1}(-V_{tM_1})}\left[\ln\left(4\frac{V_{DD}-V_{tM_2}}{V_{DD}+V_{OL}}-1\right)+\frac{1}{2}\right]$$

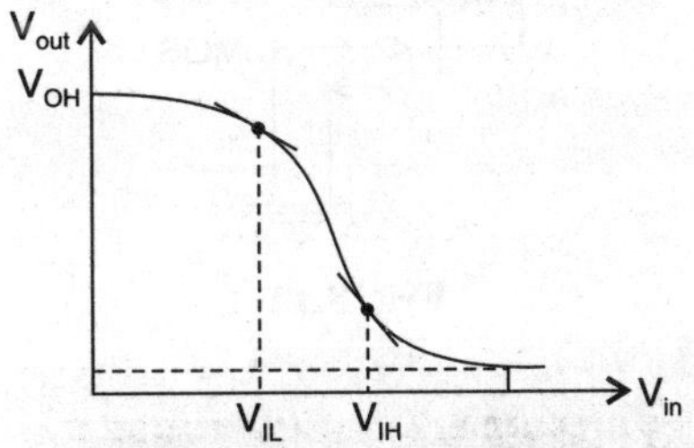

Fig. 3.10

$$\tau_{PLH} = \frac{C}{K_{M_1}(-V_{tM_1})}\left\{\ln\left[4\left(\frac{-V_{tM1}}{V_{DD}-V_{OL}}\right)-1\right]+2\frac{V_{DD}+V_{tM1}-V_{OL}}{-V_{tM1}}\right\}.$$

Rise Time :

$$t_r = \frac{C}{K_{M_1}(-V_{tM_1})}\left\{\ln\left[\left(-\frac{20\,V_{tM_1}}{\Delta V}\right)-1\right]+2\frac{V_{DD}+V_{tM_1}-V_{OL}-0.10\,V}{(-V_{tM_1})}\right\}$$

Fall Time :

$$t_f = \frac{C}{K_{M_2}(V_{DD}-V_{tM_2})}\left\{\ln\left[2\left(\frac{V_{DD}+V_{tM_2}}{V_{DD}-0.9\,\Delta V}\right)-1\right]+2\frac{V_{tM_2}-0.1\,\Delta V}{V_{DD}-V_{tM_2})}\right\}$$

Noise Margin :

$$NM_H = V_{OH} - V_{IH}$$
$$NM_L = V_{IL} - V_{OL}.$$

3.4 CMOS INVERTER

The CMOS inverter consists of an enhancement type *n*MOS and an enhancement-type *p*MOS transistor, operating in complementary mode. This configuration is called complementary MOS (CMOS).

The circuit topology is complementary purple-pull in the same that for high input, the *n*MOS transistor drives (pull down) the output mode while the *p*-MOS transistor acts as the load, and for low input the *p*-MOS transistor drives (pull-up) the output node while the *n*MOS transistor acts as the load.

Both the devices contribute equally to the circuit operation characteristics.

Advantages:

(*a*) Steady state power dissipation of the CMOS inverter circuit is virtually negligible, except for small power dissipation due to leakage current.

(*b*) VTC is very sharp.

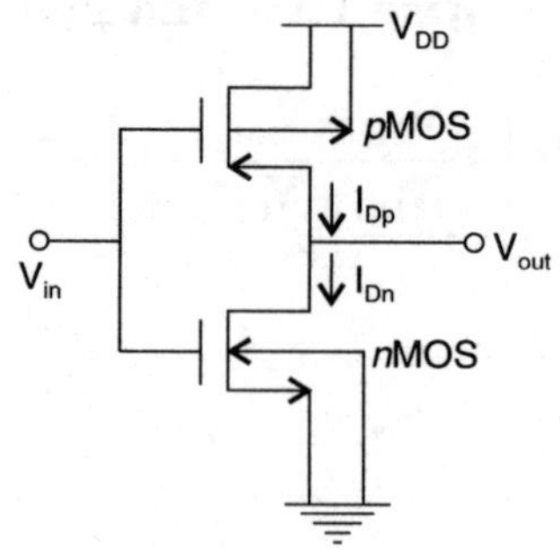

Fig. 3.11

Operation :

The input voltage is connected to the gate terminates of both the n and p MOS transistor. Thus, both of them are driven by the input signal V_{in}, directly.

It can be seen that, the S-D junction of p-MOS is reverse biased and $V_{SB,n} = 0$.

From the circuit, we can say that, for n-MOS.

$$V_{us,n} = V_{in}$$

$V_{DS,n} = V_{out}$, and for p-MOS.

$$V_{us,p} = -(V_{DD} - V_{in})$$

$$V_{DS,p} = -(V_{DD} - V_{out})$$

When, $V_{in} < V_{To,n} \rightarrow n$MOS is in cutoff region. p MOS is 'ON', operating in the linear region. Since, $I_{Dn} = I_{Dp} = 0$, then $V_{DS,p} = 0$.

$$\therefore \quad \boxed{V_{DD} = V_{out} = V_{OH}}$$

When $V_{in} > V_{DD} + V_{TO,p}$ then,

p-MOS $\rightarrow$ turned off.

n-MOS $\rightarrow$ operating in linear region.

Then, $V_{DS,n} = 0$.

$$\therefore \quad \boxed{V_{out} = V_{DS} = V_{OL} = 0}$$

When, $V_{DS,n} \geq V_{us,n} - V_{TO,n}$ then, n-MOS transistor operates in saturation region.

$$\boxed{V_{out} \geq V_{in} - V_{TO,n}}$$

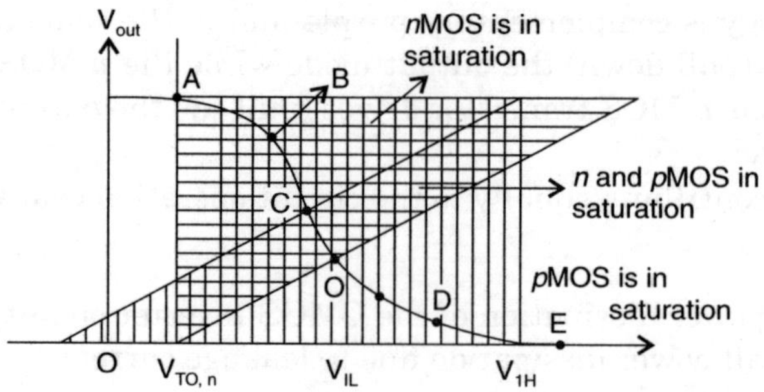

Fig. 3.12

When, $V_{in} < V_{DD} + V_{TO,p}$ and $V_{DS} \le V_{as,p} - V_{TO,p}$ then,

$$\boxed{V_{out} \le V_{in} - V_{TO,p}}$$

VTC curve of CMOS inverter operating region are described given below :

Region	V_{in}	V_{out}	*n*MOS	*p*MOS
A	$< V_{TO,n}$	V_{out}	Cutoff	Linear
B	V_{IL}	$\approx V_{out}$	Sat	Linear
C	V_{in}	V_{in}	Sat	Sat
D	V_{IH}	$\approx V_{OL}$	Linear	Sat
E	$> V_{DD} + V_{TO,p}$	V_{OL}	Linear	Cutoff

Calculation of V_{IL} :

At $V_{in} = V_{IL}$, then $\dfrac{dV_{out}}{dV_{in}} = -1$.

*n*MOS operates in saturation region and *p*MOS operates in linear region.

From, $I_{D,n} = I_{D,p}$.

$$\frac{K_n}{2}(V_{as,n} - V_{TO,n})^2 = \frac{K_p}{2}[2(V_{us,p} - V_{TO,p})V_{DS,p} - V_{DS}{}^2{}_{,p}]$$

or, $$\frac{K_n}{2}(V_{in} - V_{TO,n})^2 = \frac{K_p}{2}[2(V_{in} - V_{DD} - V_{TO,p})(V_{out} - V_{DD}) - (V_{out} - V_{DD})^2]$$

$[\because V_{DS,p} = -(V_{DD} - V_{out})$ and $V_{gs,p} = -(V_{DD} - V_{in})]$

and $[V_{gs,n} = V_{in} = V_{DS,n} = V_{out}]$.

Differentiate both side w.r.t. V_{in}, we get,

$$K_n(V_{in} - V_{TO,n}) = K_p\left[(V_{in} - V_{DD} - V_{TO,p})\left(\frac{dV_{out}}{dV_{in}}\right) + (V_{out} - V_{DD}) - (V_{out} - V_{DD})\left(\frac{dV_{out}}{dV_{in}}\right)\right]$$

Substituting, $V_{in} = V_{IL}$ and $\dfrac{dV_{out}}{dV_{in}} = -1$ we get,

$$K_n(V_{IL} - V_{TO,n}) = K_p(2V_{out} - V_{IL} + V_{TO,p} - V_{DD})$$

$$V_{IL} = \frac{2V_{out} + V_{TO,p} - V_{DD} + K_R + V_{TO,n}}{1 + K_R}$$

where $K_R = \dfrac{K_n}{K_p}$.

Calculation of V_{IH} :

Here, $\frac{dV_{out}}{dV_{in}} = -1$ and $V_{in} = V_{IH}$.

∴ nMOS is in linear region and pMOS operation saturation region.

∴ $I_{O,n} = I_{D,p}$.

$$\therefore \quad \frac{K_n}{2} \cdot [2(V_{gs\,in} - V_{TO,n}) V_{DS,n} - V_{DS}{}^2{}_{,n}]$$

$$= \frac{K_p}{2} (V_{gs,p} - V_{TO,p})^2 = \frac{K_p}{2}$$

Substituting, $V_{in} = V_{IH}$ and $\frac{dV_{out}}{dV_{in}} = -1$, we get,

$$K_n(-V_{IH} + V_{TO,n} + 2V_{out}) = K_p(V_{IH} - V_{DD} - V_{TO,p})$$

$$\therefore \quad V_{IH} = \frac{V_{DD} + V_{TO,p} + K_R(2V_{out} + V_{TO,n})}{1 + K_R}.$$

Calculation of V_{th} :

The inverter threshold voltage is defined as,

$$V_{th} = V_{in} = V_{out}.$$

For $V_{in} = V_{out}$, both transistors are operated in saturation region. So,

$$I_{D,n} = I_{D,p}.$$

$$\therefore \quad \frac{K_n}{2}(V_{gs,n} - V_{TO,n})^2 = \frac{K_p}{2}(V_{gs,p} - V_{TO,p})^2$$

$$\text{or,} \quad \frac{K_n}{2}(V_{in} - V_{TO,n})^2 = \frac{K_p}{2}(V_{in} - V_{DD} - V_{TO,p})^2$$

$$\text{or,} \quad V_{in}\left(1 + \sqrt{\frac{K_p}{K_n}}\right) = V_{TO,n} + \sqrt{\frac{K_p}{K_n}}(V_{DD} + V_{TO,p})$$

Switching threshold voltage,

$$V_{th} = \frac{V_{TO,n} + \sqrt{\frac{K_p}{K_n}}(V_{DD} + V_{TO,p})}{1 + \sqrt{\frac{K_p}{K_n}}}$$

$$= \frac{V_{TO,n} + \sqrt{\frac{1}{K_R}}(V_{DD} + V_{TO,p})}{1 + \sqrt{\frac{1}{K_R}}}$$

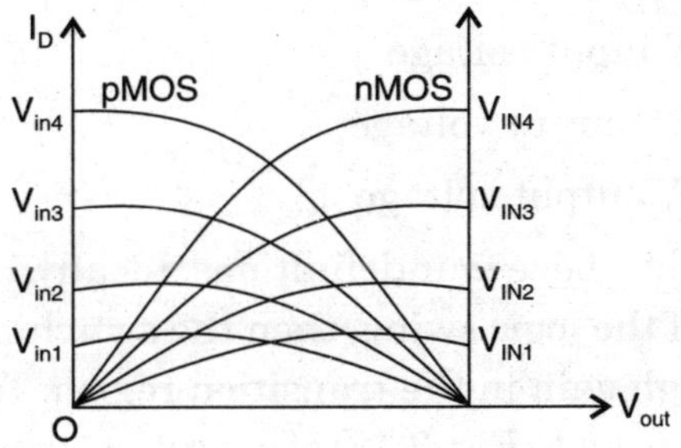

Fig. 3.13

I-V characteristics of CMOS inverter.

Noise margin : is a parameter intimately related to the transfer characteristics. It allows one to estimate the allowable noise voltage on the input of a gate so that the output will not be affected. Noise margin (also called noise immunity) is specified in terms of two parameters – the low noise margin NM_L, and the high noise margin NM_H. Referring to Figure below, NM_1 is defined as the difference in magnitude between the maximum LOW input voltage recognized by the driven gate and the maximum LOW output voltage of the driving gate. That is,

$$NM_L = | V_{ILmax} - V_{OLmax} |$$

Similarly, the value of NM_H is the difference in magnitude between the minimum HIGH output voltage of the driving gate and the minimum HIGH input voltage recognizable by the driven gate. That is,

$$NM_H = | V_{OHmin} - V_{IHmin} |$$

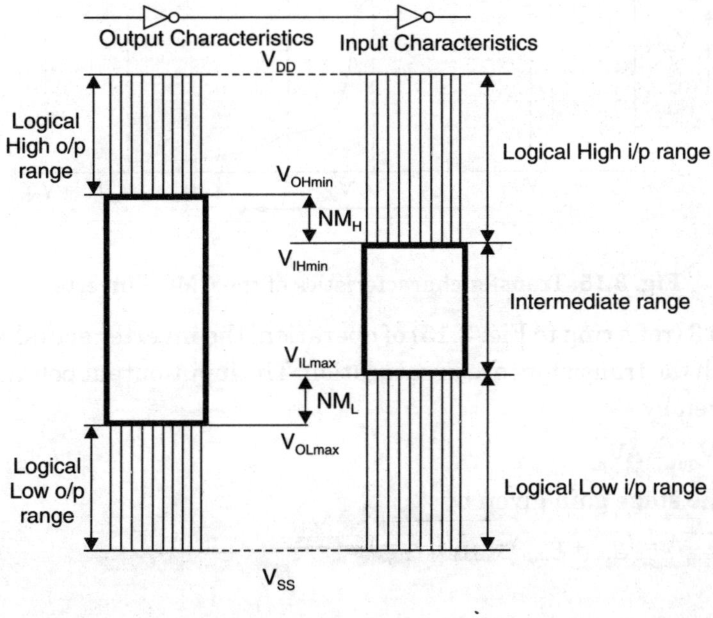

Fig. 3.14. Definition of noise margin.

where V_{IHmin} : minimum HIGH input voltage

$V_{IL\,max}$: maximum LOW input voltage

V_{OHmin} : minimum HIGH output voltage

V_{OHmax} : maximum LOW output voltage

Figure 3.14 illustrates the above four definitions. Ideally, if one desires to have $V_{IH} = V_{IL}$, and $V_{OL} = V_{OH}$ in the middle of the logic swing, then the switching of states should be abtrupt, which in turn requires very high gain in the transition region. To calculate V_{IL}, the inverter is supposed to be in region 2 (referring to Fig. 3.12) of operation, where the *p*-transistor is in linear zone while the *n*-transistor is in saturation. The parameter V_{IL} is found out by considering the unity gain point on the inverter transfer characteristics where the output makes a transition from V_{OH}. Similarly, the parameter V_{IH} is found by considering the unity gain point at the V_{OL} end of the characteristics.

If the noise margins NM_H or NM_L are reduced to a low value, then the gate may be susceptible to switching noise that may be present at the inputs. The net effect of noise sources and noise margins on cascaded gates must be considered in estimating the overall noise immunity of a particular system. Not infrequently, noise margins are compromised to improve speed.

3.5 CMOS INVERTER AS AN AMPLIFIER

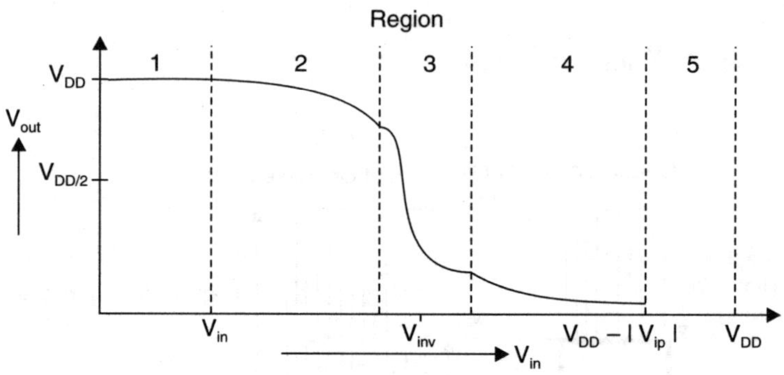

Fig. 3.15. Transfer characteristics of the CMOS inverter

In the region 3 (referring to Fig. 3.15) of operation, the inverter actually acts as an analog amplifier where both the transistors are in saturation. The input-output behaviour of the inverter in this region is given by

$$V_{out} = AV_{in}$$

where A is the stage gain given by

$$A = (g_m + g_{mp})\,(r_{dsn} \parallel r_{dsp})$$

Note that the small-signal characteristics, namely transconductance g_m is defined as

$$g_m = \frac{dI_{ds}}{dV_{gs}} \qquad | \, V_{ds} : \text{some fixed value}$$

and the output resistance r_{ds} is given by

$$r_{ds} = \frac{dV_{ds}}{dI_{ds}} \qquad | \, V_{ds} : \text{some fixed value}$$

Note that gain A is dependent on the process and the transistors used in the circuit. It can be increased by increasing the length of the transistors to improve the output resistance. However, speed and bandwidth of the amplifier suffer as a result.

Chapter 4 STATIC VLSI DESIGN

4.1 INTRODUCTION

There are numerous circuit styles to implement a given logic function. As with the inverter, the common design metrics by which a gate is evaluated are area, speed, energy and power. Depending on the application, the emphasis will be on different metrics, by which a gate is evaluated. For example, the switching speed of digital circuits is the primary metric in a high performance processor, while in a battery operated circuit, it is energy dissipation.

Recently, power dissipation also has become an important concern and considerable emphasis is placed on understanding the sources of power and approaches to dealing with power. In addition to there metrics, roburtness to noise and reliability are also very important considerations.

The complementary CMOS circuit style falls under a broad class of logic circuits called static circuits in which at every point in time, each gate output is connected to either V_{DD} or V_{SS} via a low resistance path.

There are different static design procedure exist in VLSI circuit design technology:

(*a*) Complementary CMOS design (pull-up and pull-down network design technology)

(*b*) Depletion load logic style

(*c*) Ratioed logic style (Pseudo-*n*MOS logic style)

(*d*) Pass transistor logic style

(*e*) Transmission logic style

(*f*) DCVSL logic style.

4.2 PULL-UP AND PULL-DOWN NETWORK DESIGN STYLE

The function of the PUN (pull-up network) is to provide a connection between the output and V_{DD} anytime the output of the logic gate is meant to be1 (based on the inputs).

Similarly, the function of PDN (pull-down network) is to connect the output of V_{SS} when the output of the logic gate is meant to be '0' (based on the inputs).

The PUN and PDN networks are constructed in a mutually exclusive fashion such that one and only one of the networks is conducting in steady state.

When, once the transients have settled, a path always exists between V_{DD} and the output F for a high output ("one"), or between V_{ss} and F for a low output ("zero"). This is equivalent to stating that the output node is always a low-impedance node in steady state.

Design Consideration:

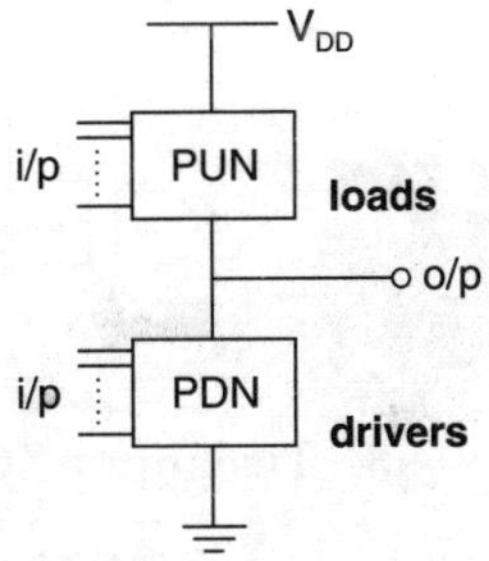

- An NMOS switch is 'on' when the controlling signal (at the o/p at the gate) is high and is 'off' when the controlling signal is low.
- A PMOS transistor acts as an inverse switch that is 'ON' when the controlling signal is 'low' and 'OFF' when the controlling signal is 'high'.
- The PDN is constructed using NMOS devices, while PUN is constructed using PMOS.
- The primary reason for this choice is that NMOS transistors, produce "strong zero", and PMOS transistors produce "strong ones".
- An NMOS device pulls the o/p all the way down to GND, while a PMOS lowers the o/p no further than $|V_{TP}|$ – the PMOS 'turns off' at that point and stops contributing discharge current. For this reason the NMOS are thus preferred devices in PDN.
- A PMOS switch succeeds in clearging the o/p all the way to V_{DD}, while the NMOS device fails to raise the output above $V_{DD} - V_{Tn}$. For this reason, the PMOS are we in the PUN.
- The OR operation is performed by parallel — connected drivers and series connected loads.
- The AND operation is performed by series connected drivers and parallel connected load.
- The inversion is performed at the point. Where the load and the drivers meet which is the nature of MOS circuit operation.
- There above state design principle can also be extended to circuit sub-blocks so that Boolean OR and AND operations can be performed in a nested circuit structure.

Examples:

1. Two input NAND Gate :

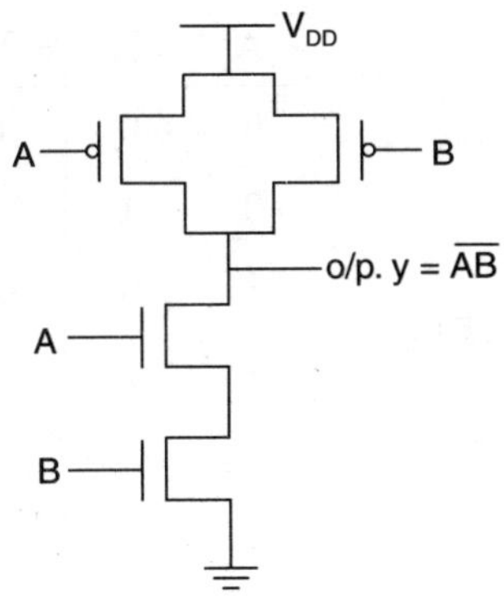

A	B	Y
0	0	1
0	1	1
1	0	1
1	1	0

2. Two input NOR Gate:

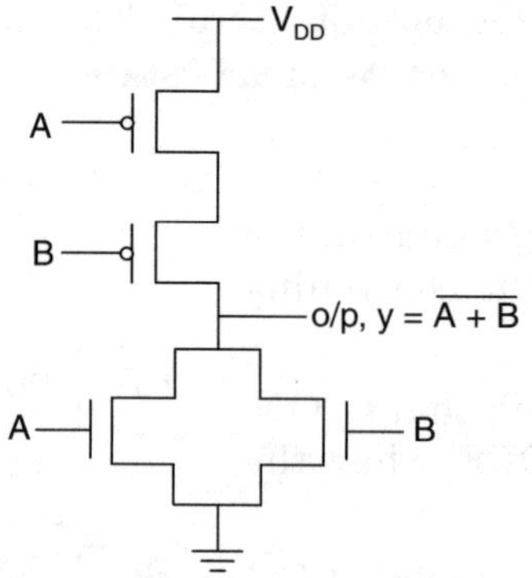

A	B	Y
0	0	1
0	1	0
1	0	0
1	1	0

3. Two input XOR Gate:

$$Y = \overline{A}B + A\overline{B} = \overline{AB + \overline{A}\,\overline{B}}$$

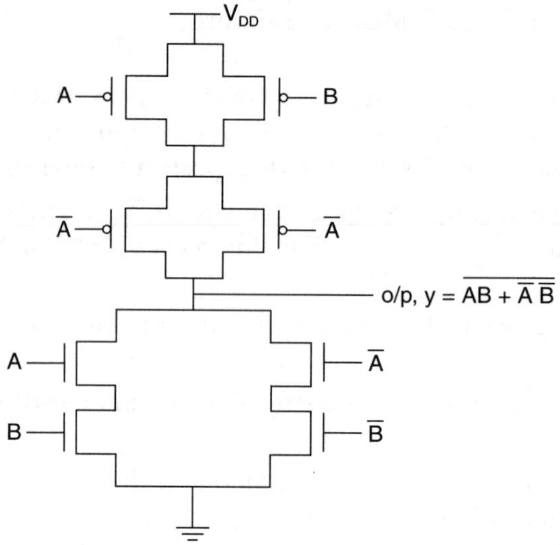

A	B	Y
0	0	0
0	1	1
1	0	1
1	1	0

4. The two input XNOR Gate:

$$Y = AB + \overline{A}\,\overline{B} = \overline{\overline{A}B + A\overline{B}}$$

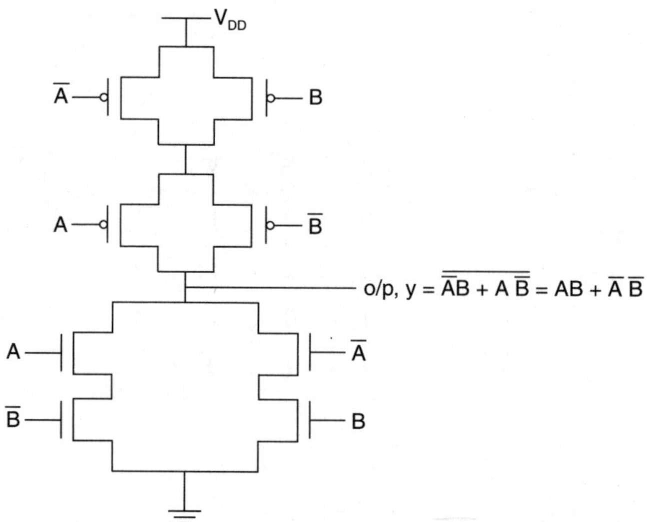

A	B	Y
0	0	1
0	1	0
1	0	0
1	1	1

5. Design : $F = \overline{D + A(B + C)}$

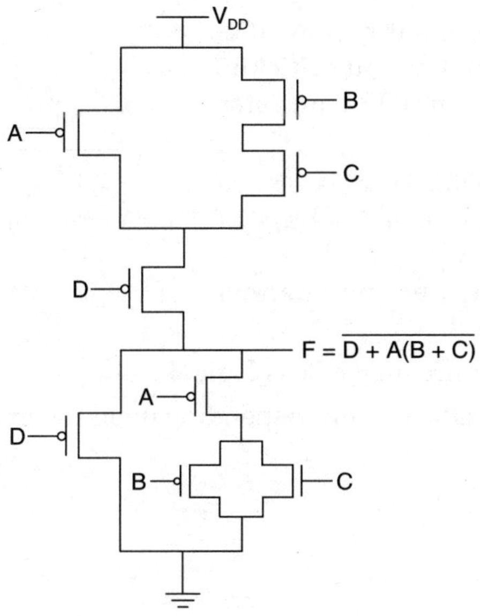

6. $Y = \overline{(A + B + C)(D + E)F}$

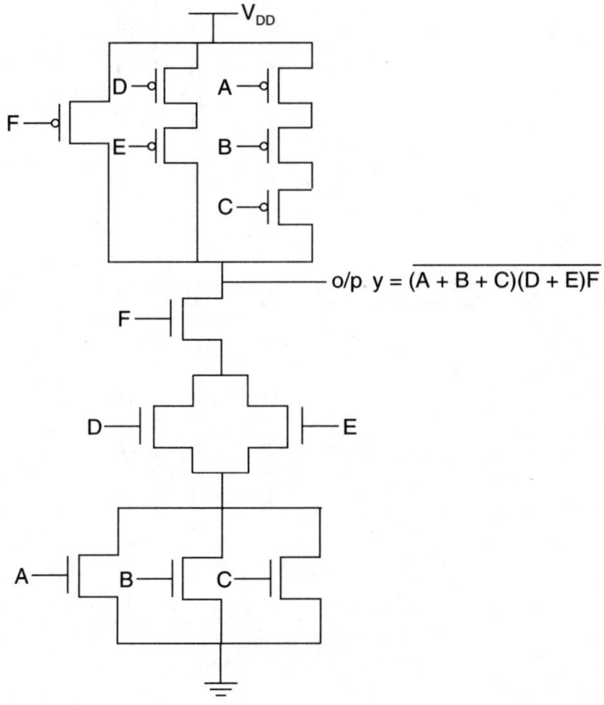

4.3 DEPLETION LOAD LOGIC STYLE

The circuit diagram of the depletion load circuit is shown in figure below.

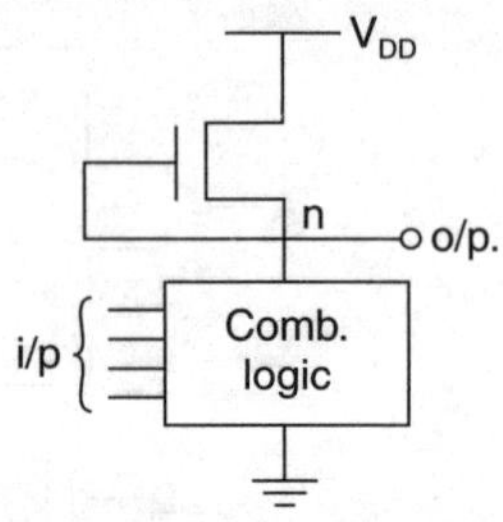

The driver network is built by the enhancement type NMOS type transistor depending upon the combinational circuit with the $V_{TO,\ driver} > 0$ whereas the load is a depletion – type *n*MOS transistor with $V_{TO,\ load} < 0$.

The gate and the source nodes of the load transistor are connected ; hence, $V_{as,\ load} = 0$ always. Since the threshold voltage of the depletion load is negative, the condition, $V_{as,\ load} > V_{TO,\ load}$ is satisfied and the load device always has a conducting channel regardless of the input and output voltage levels.

So, the voltage at node 'P', shown in the figure above is as logic '1'.

This voltage is properly evaluated by the PDN network depending upon the proper input voltages.

Examples.

1. Two input NAND Gate:

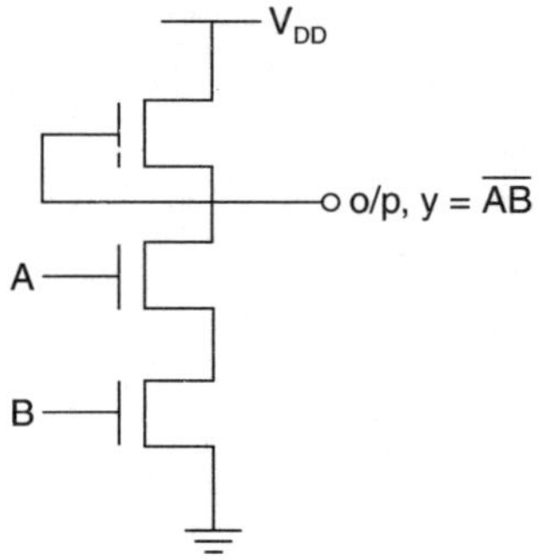

A	B	Y
0	0	1
0	1	1
1	0	1
1	1	0

2. Two input NOR Gate:

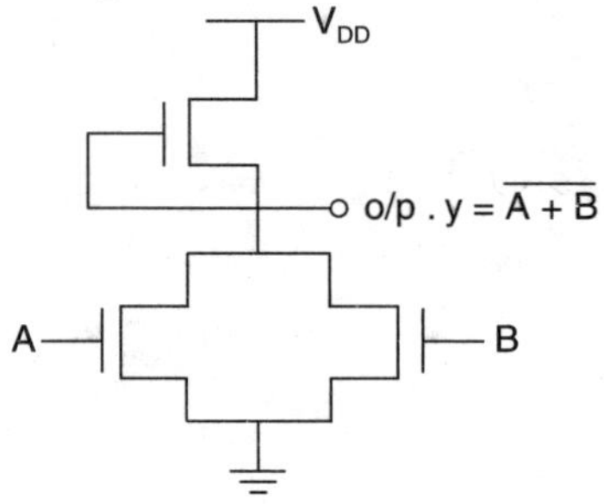

A	B	Y
0	0	1
0	1	0
1	0	0
1	1	0

3. Two input XOR Gate:

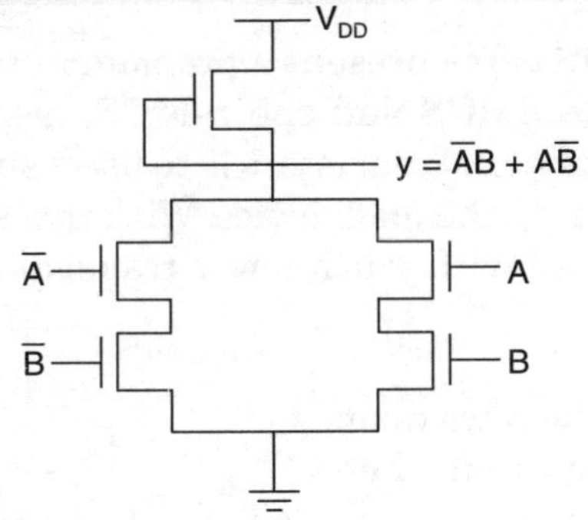

$$Y = \overline{\overline{A}\,\overline{B} + AB} = \overline{A}\,B + A\overline{B}$$

A	B	Y
0	0	0
0	1	1
1	0	1
1	1	0

4. Two input XNOR Gate:

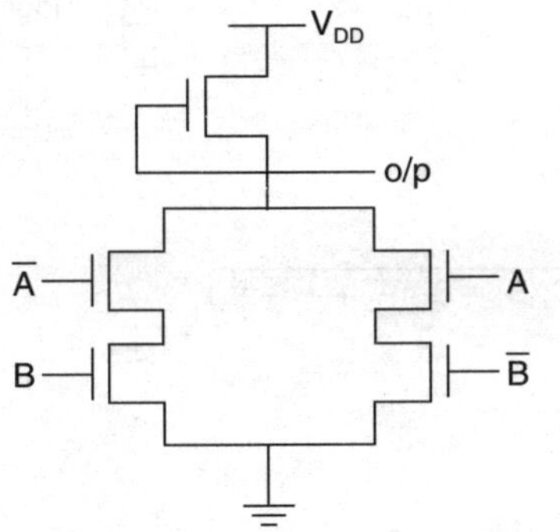

$$Y = \overline{\overline{A}\,B + A\,\overline{B}} = AB + \overline{A}\,\overline{B}$$

5. $F = \overline{A(B + C) + D}$

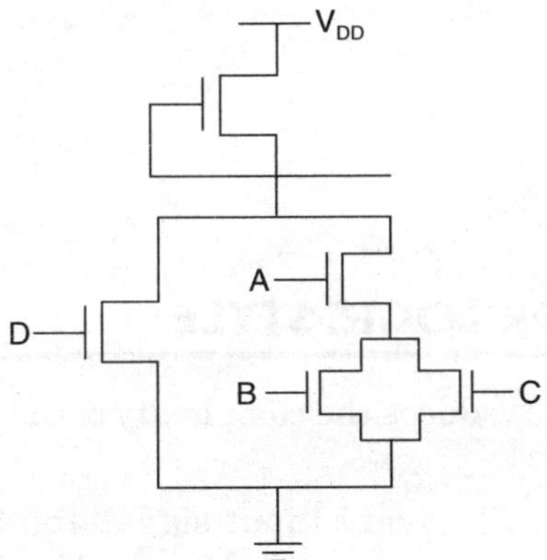

Advantages of this type design:

(*i*) Single power supply required.

(*ii*) Better noise margin.

(*iii*) Smaller overall layout area.

4.4 RATIOED LOGIC STYLE

The large area requirements of complex CMOS gates present a problem in high density designs, service two complementary transistors, one *n*MOS and one *p*-MOS, are needed for every input. One possible approach to reduce the number of transistors is to use a single*p*-MOS transistor, with its gate terminal connected to ground, as the load device. With this simple pull-up arrangement, the complex gate can be implemented with much fewer transistor.

Disadvantage:

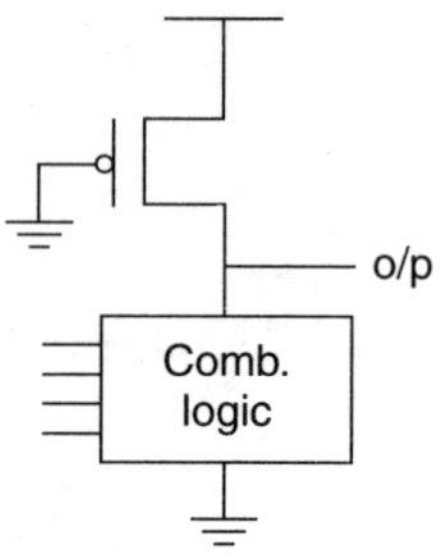

1. Non-zero static power dissipation, since the always on *p*-MOS load device conducts a steady state current when the o/p voltage is lower than V_{DD}.

Example:

$$F = \overline{C_1(B_1 + B_2)(A_1 + A_2 + A_3)}$$

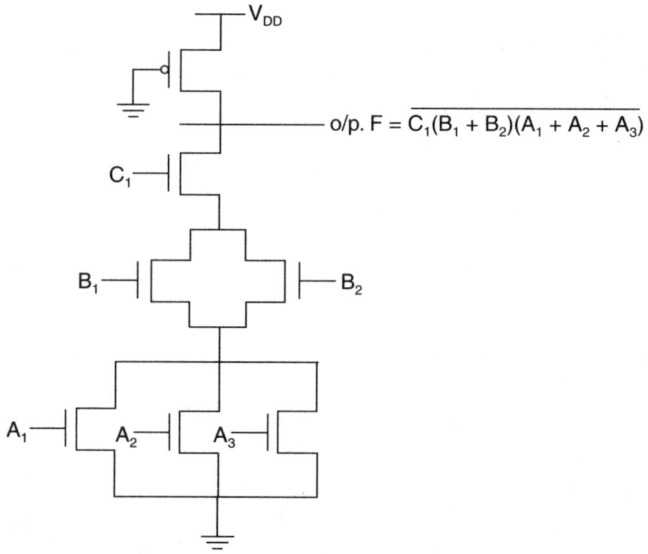

4.5 COMPLEMENTARY PASS TRANSISTOR LOGIC STYLE

The Complementary Pass Transistor Logic (CPL) reduces the complexity of the full CMOS pass gate logic circuit.

All inputs are applied in complementary form, *i.e.,* every input signal and its inverse must be provided and the circuit also produces complementary outputs to be used by subsequent CPL states.

Thus, the CPL circuit essentially consists of complementary inputs, an *n*MOS pass transistor logic network to generate complementary outputs, and CMOS output inverters to restore the output signals.

The elimination of*p*MOS transistors for the pass gate reduces the parasitic capacitances. So high speed of operation is achieved. CPL design style is highly modular and wide range of functions can be realized by using the same basic pass transistor structures.

Disadvantages:

To eliminate threshold voltage drops, the threshold voltage of the nMOS transistors in the pass gate network must be reduced to zero. This reduces the overall noise immunity and the problems like subthreshold conduction in off mode occurs.

Examples:

1. NAND/AND:

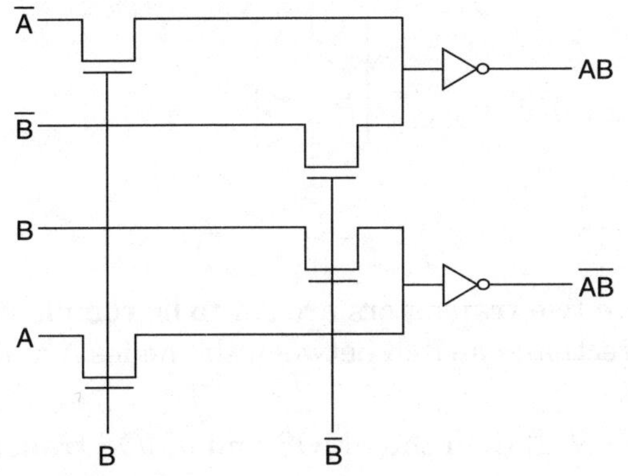

2. NOR/OR:

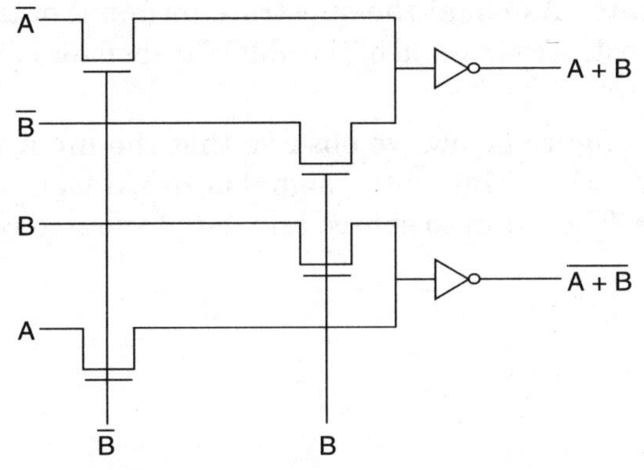

3. XNOR/XOR:

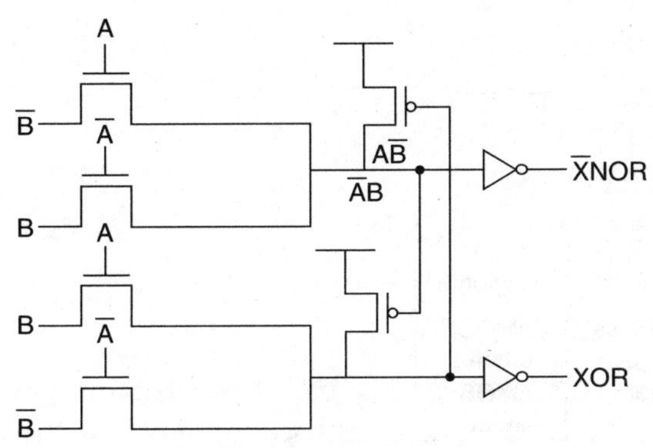

4.6 TRANSMISSION GATE LOGIC STYLE

A CMOS transmission gate consists of one *n*MOS and one *p*MOS transistor and they are connected in parallel. The representation of the CMOS transmission gate (TG) is shown in figure given below:

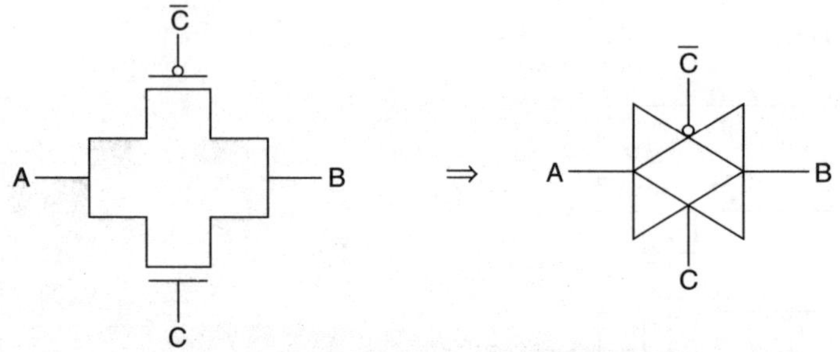

The voltage applied to the gate of there two transistors are set to be complementary of each other. The CMOS TG operates as a bidirectional switch between the nodes A and B and is controlled by signal *c*.

If the control signal *c* is at logic high (= V_{DD}) then the *p*MOS and *n*MOS transistors are turned on and they provide a low resistance current path between A and B.

If the control signal *c* is low then both the transistors are off and there is no conduction path between A and B (high impedance state). As usual the substrate terminal of the *n*MOS transistor is connected to ground and the substrate terminal of the *p*MOS transistor is connected to V_{DD}.

From the bias condition given in the figure below, we observe that the input node A is connected to a constant logic high voltage $V_{in} = V_{DD}$. The control signal is also at logic high level. So the both transistors are in the ON state. The drain to source and gate to source voltages of the *n*MOS transistors are

$$V_{DS,\,n} = V_{DD} - V_{out}$$
$$V_{as,\,n} = V_{DD} - V_{out}$$

Biasing of TG.

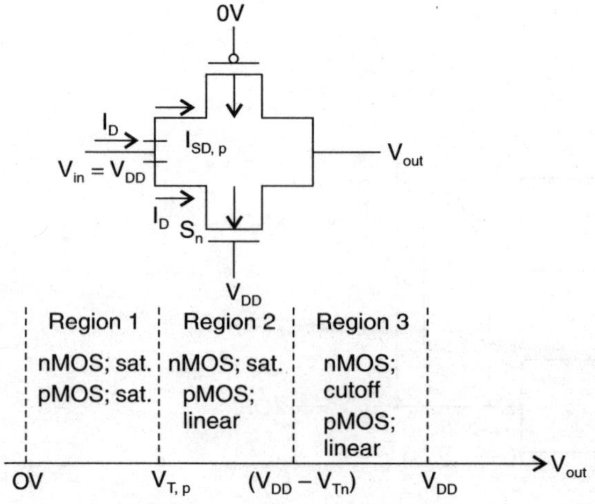

Region of operation of TG. The nMOS transistor will be turned off for $V_{out} > V_{DD} - V_{T,n}$ and will operate in the saturation region for $V_{out} < V_{DD} - V_{T,n}$. The drain to source and gate to source voltage of the pMOS transistor is given as,

$$V_{DS,p} = V_{out} - V_{DD}$$
$$V_{as,p} = -V_{DD}$$

The pMOS transistor is in saturation for $V_{out} < |V_{TP}|$ and it is in linear region for $V_{out} > |V_{TP}|$. The pMOS transistor is always turned on irrespective of the output voltage V_{out} level.

The operating regions changed with the change is the o/p voltage level V_{out}.

The total current flowing through the TG in the sum of the nMOS and pMOS drain current. The equivalent resistances of the transistors is given by

$$R_{eq,n} = \frac{V_{DD} - V_{out}}{I_{DS,n}}$$

$$R_{eq,p} = \frac{V_{DD} - V_{out}}{I_{SD,p}}$$

The total equivalent resistance is given by,

$$R_{tot} = R_{eq,n} \parallel R_{eq,p}$$

The total equivalent resistance of the CMOS TG remains almost constant. It has a very little variation with the change in output voltage. But the individual equivalent resistances are strongly dependent on the output voltage.

So, we see that a CMOS TG is turned on by a logic high. Control signal and can be used in logic circuit design.

Example:

1. Design a 2 × 1 MUX :

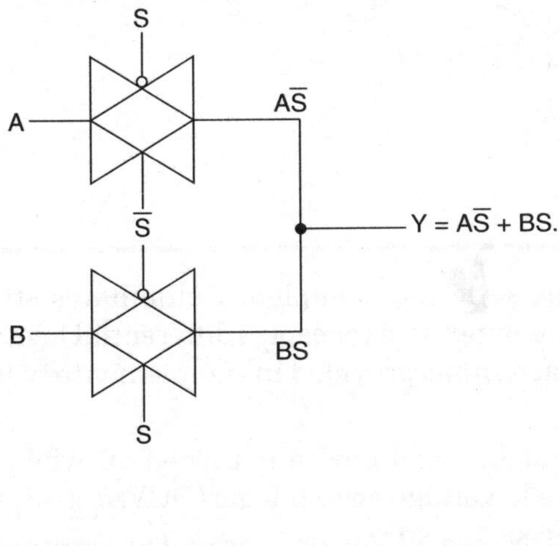

$Y = A\bar{S} + BS$

S	Y
0	A
1	B

2. Design an XOR Gate :

$Y = A\overline{B} + \overline{A}B$

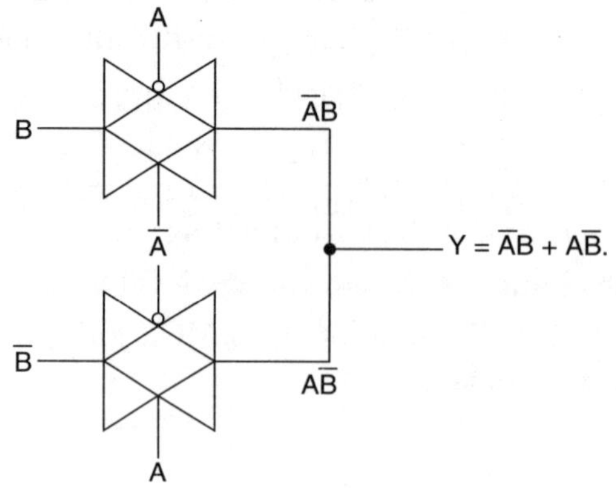

3. $F = AB + \overline{A}\ \overline{C} + A\overline{B}C = A(B + \overline{B}C) + \overline{A}\ \overline{C}$

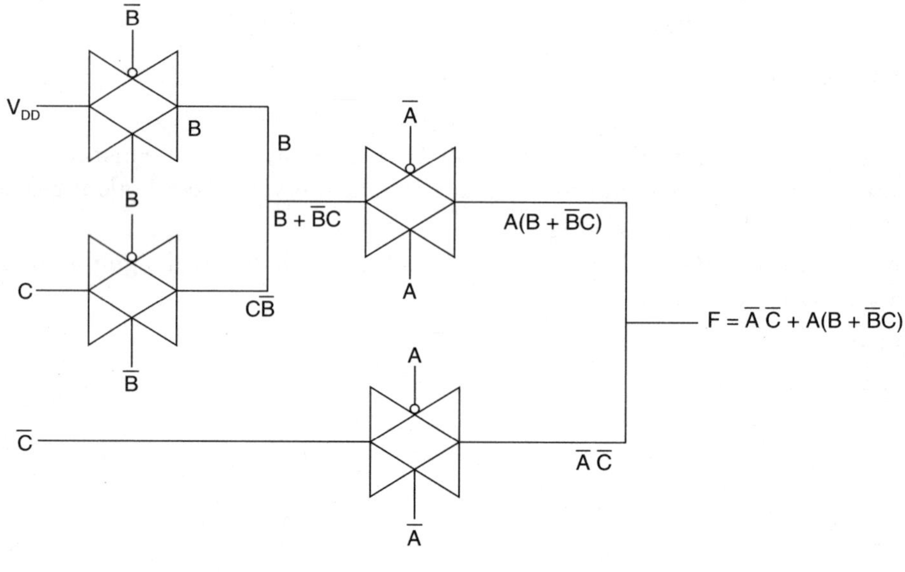

4.7 DCVSL LOGIC GATE

It is possible to create a rational logic style that completely eliminates static currents and produces rail-to-rail swing. Such a gate combines two concepts : differential logic and positive feedback. A differential gate requires that each input provided in complementary format and it produces complementary output.

The feedback mechanism ensures that the load device is turned off when not needed. Such logic family is called. Differential cascode voltage switch logic (DCVSL).

The pull down networks PDN1 and PDN2 we NMOS devices and are mutually exclusive that is, when PDN1 conducts, PDN2 is off and when PDN1 is off, PDN2 conducts — such that the required logic function and its inverse are simultaneously implemented.

Assume now that, for a given set of inputs, PDN1 conducts while PDN2 does not, and that out and $\overline{out}$ are initially 'high' and 'low' respectively.

Turning 'on' PDN1, causes out to be pulled down, although there is still contention between M_1 and PDN1. $\overline{out}$ is in a high impedance state, as M_2 and PDN2 are both turned off. PDN1 must be strong enough to bring 'out' below $V_{DD} - |V_{TP}|$, the point at which M_2 turns 'on' and starts charging $\overline{out}$ to V_{DD}, eventually turning off M_1. This in turn enables out to discharge all the way to GND.

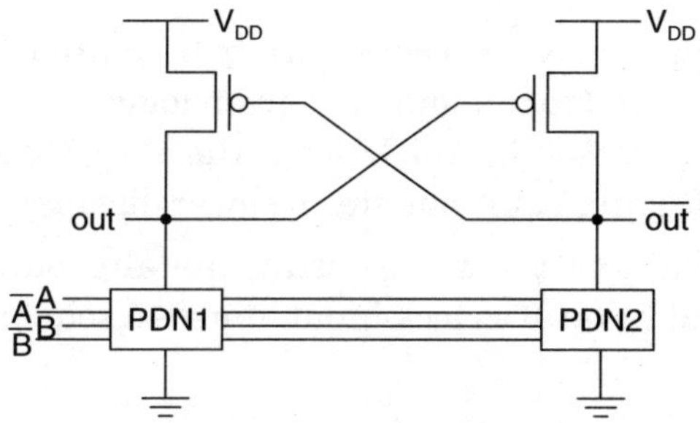

Basic Principle

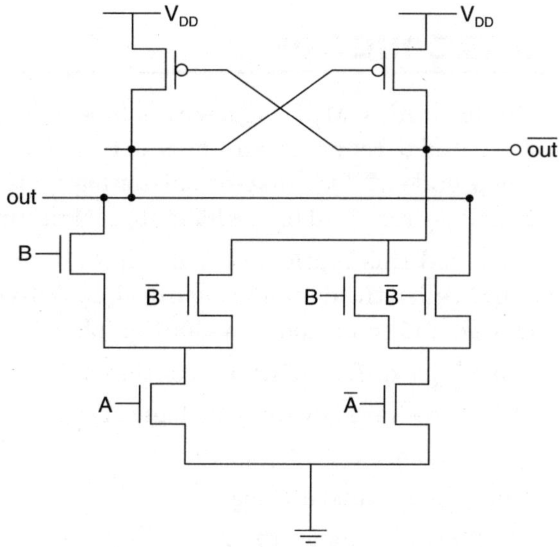

$A\overline{B} + \overline{A}B$

XOR — XNOR circuit.

Chapter 5

DYNAMIC VLSI DESIGN

5.1 INTRODUCTION

In the static CMOS logic with a fan-in of N requires 2N devices. A variety of approaches were presented to reduce the number of transistors required to implement a given logic function including pseudo-*n*MOS, pass-transistor logic. etc. For the pseudo, *n*MOS logic, there are (N + 1) no of transistor required to implement an N input logic gate, but it has static power dissipation.

To avoid this static power dissipation and obtaining the same result, the alternating approach is called the dynamic logic design. With the addition of a clock input, it uses a sequence of precharge and conditional evaluation phase.

There are different style to design a dynamic VLSI circuits:

(*i*) Precharge evaluation logic style.

(*ii*) Dynamic TG logic style.

(*iii*) Pass transistor logic style.

(*iv*) Domino logic style.

(*v*) NORA logic style.

5.2 PRE-CHARGE AND EVALUATION LOGIC STYLE

The operation of this circuit is divided into two major phases — pre-charge and evaluation — with the mode of operation determined by the clk signal clk. The basic construction of this type of design style is given below.

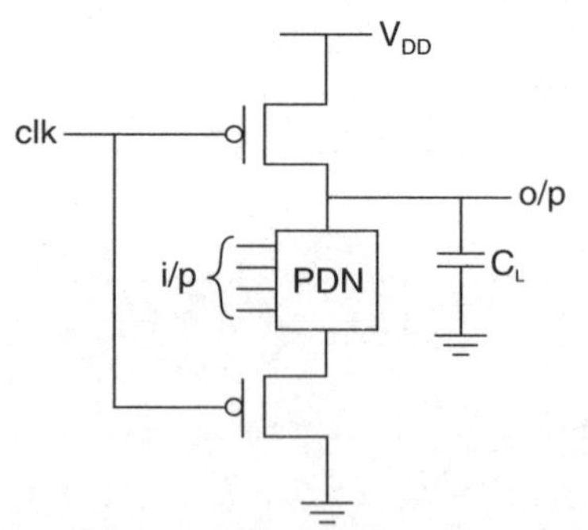

The PDN network is built by the *n*MOS with respect to the combinational inputs.

Pre-charge:

When clk = 0, the output node 'out' is pre-charged to V_{DD} by the *p*MOS transistor (M_p). During that time, the evaluation transistor *n*MOS (M_e) is off, so that the pull down path is disabled. The evaluation of FET elimination of any static power that would be consumed during the pre-charge period because static. Current would flow between the supplies if both the pre-charge device and PDN were turned on simultaneously.

Evaluation:

When clk = 1, the M_p is off and M_e is turned on. The o/p is conditionally discharged based on the input values and the pull down topology.

If the inputs are such that the PDN conducts, then a low resistance path exists between out and GND, and the output is discharged to GND.

If the PDN is turned off, the precharged value remains stored on the output capacitance C_L, which is a combination of junction capacitances, the wiring capacitance and the input capacitance of the fan-out gates.

During the evaluation phase, the only possible path between the output node and a supply rail is to GND. Consequently, once 'out' is discharged, it cannot be charged again until the next precharge operation.

The inputs to the gate can thus make at most one transition during evaluation.

Examples:

$F = \overline{A . B + C}$

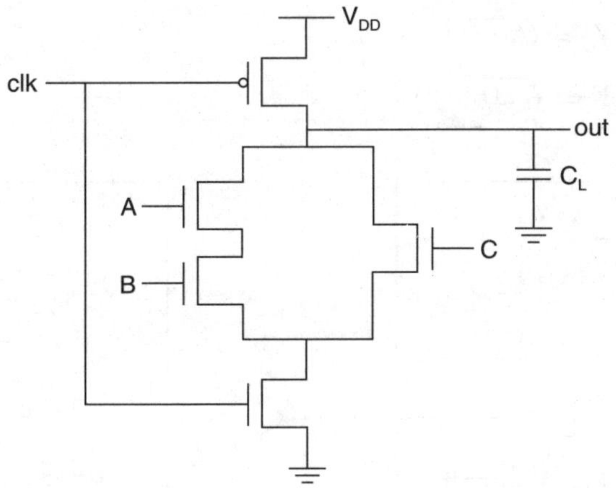

5.3 CMOS DYNAMIC TG LOGIC

In this kind of logic circuit design the static CMOS gates the used for implementing the logic blocks and CMOS transmission gates are used for transferring the output levels of one stage to the inputs of next stage.

A simple implementation of a dynamic CMOS TG logic is shown in figure given below.

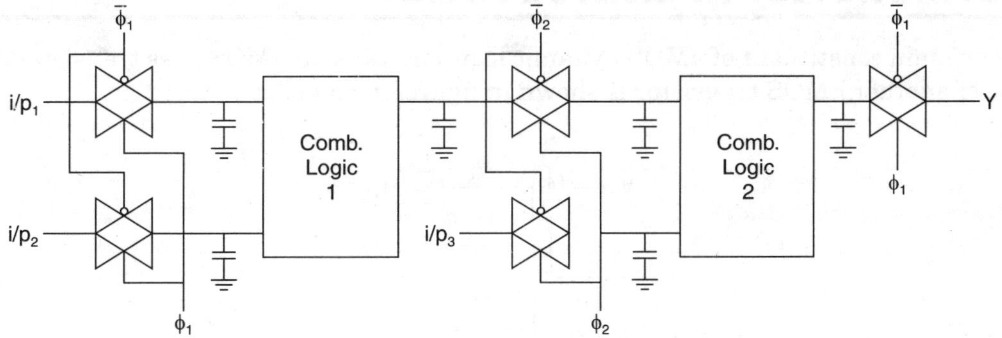

It can be seen that each transmission gate is actually controlled by the clock signal and its complement. So, with this phase clocking in CMOS transmission gate, a total of four clock signals (ϕ_1, $\overline{\phi_1}$, ϕ_2, $\overline{\phi_2}$) are required.

Here in inactive clock cycles charge is stored, in the parasitic input capacitances. During active ϕ_1 the inputs i/p_1, and i/p_2 are transferred to the input capacitances of the stage 1 and a certain output level is achieved which is stored in output capacitor.

When ϕ_2 is active this is transferred to the i/p capacitance of stage 2 through the *c*MOS TG.

The low on resistance of the CMOS transmission gate results in a smaller transfer time compared to these for *n*MOS only switches.

There is no threshold voltage drop across the CMOS TG.

Example :

$$F = \overline{(A+B)C \cdot D}$$

Output of stage 1 : $Y_1 = \overline{(A+B)C}$

Output of stage 2 : $F = \overline{Y_1 \cdot D}$

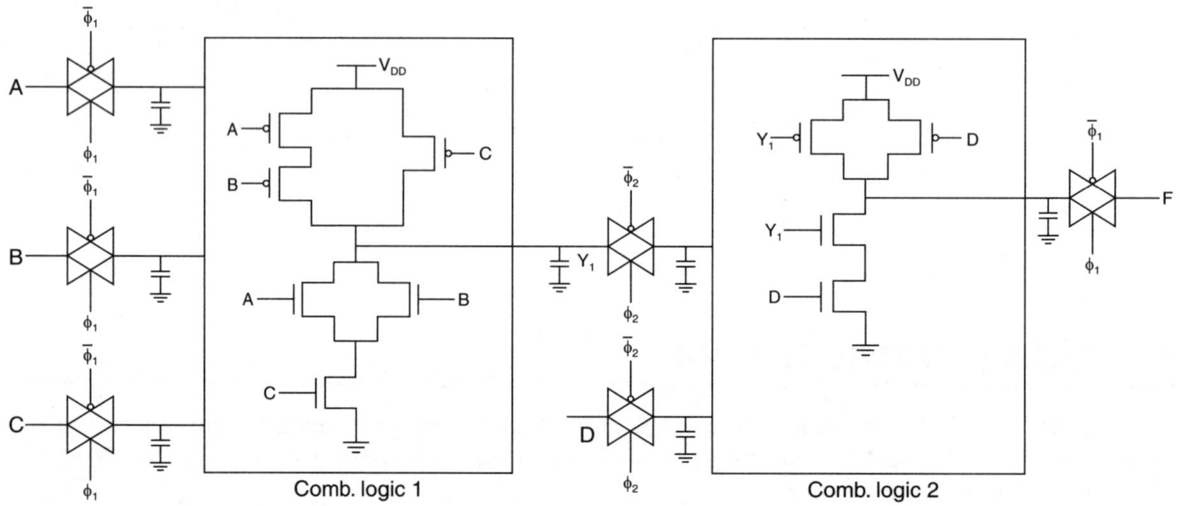

5.4 DYNAMIC PASS TRANSISTOR LOGIC

The main constituent of *n*MOS dynamic logic circuit is an *n*MOS pass transistor driving the gate of another *n*MOS transistor is shown in figure given below :

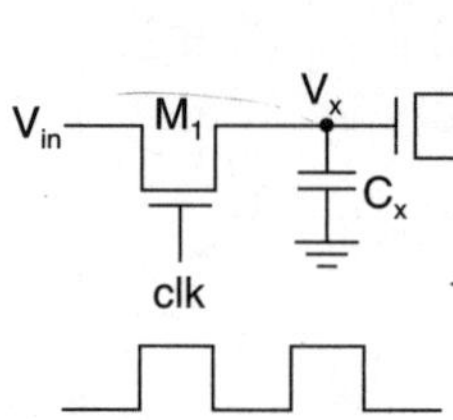

The pass transistor M_1 is driven by periodic clock signal clk and can either charge or discharge the capacitor C_X depending on input V_{in}. The pass transistor M_1 is ON only when clk = 1. When clk = 0 the *n*MOS pass transistor is off. So for clk = 1, we have two possible operations :

(*a*) **Logic 1 transfer:** Transistor M_1 provides the current path to the intermediate capacitive mode V_X. Assuming that at this node voltage V_X (t = 0) = OV initially. If V_{in} = logic 1, then $V_{in} = V_{OH} = V_{DD}$. Since the clock signal is active, M_1 is conducting. Here M_1 operates in saturation throughout this cycle since $V_{DS} = V_{as}$ and $V_{DS} > V_{as} - V_{T,n}$. The equivalent circuit of logic 1 transfer is shown in given below :

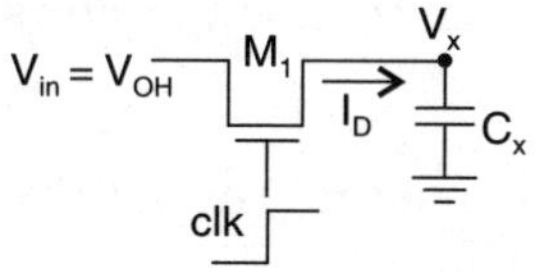

Since M_1 is conducting in saturation and the saturation drain current starts to charge the capacitor C_X. So one can write that,

$$C_X \frac{dV_X}{dt} = \frac{K_n}{2}(V_{as} - V_{T,n})^2 = \frac{K_n}{2}(V_{DD} - V_X - V_{Tn})^2 \qquad ...(i)$$

Neglecting the surface bias effect and integrating eqn. (*i*), then we get,

$$\int_0^t dt = \frac{2C_X}{K_n}\int_0^{V_X} \frac{dV_X}{(V_{DD} - V_X - V_{T,n})^2}$$

$$\therefore \quad t = \frac{2C_X}{K_n}\left[\left(\frac{1}{V_{DD} - V_X - V_{T,n}}\right) - \left(\frac{1}{V_{DD} - V_{T,n}}\right)\right] \qquad ...(ii)$$

Writing eqn. (*ii*) in terms of V_X we have,

$$V_X(t) = (V_{DD} - V_{T,n}) \frac{\left\{\frac{K_n(V_{DD} - V_{T,n})}{2C_X}\right\}t}{1 + \left\{\frac{K_n(V_{DD} - V_{T,n})}{2C_X}\right\}t} \qquad ...(iii)$$

From the eqn. (*iii*) we can say that, $V_X(t)$ grows with time with a maximum of $V_{X(max)} = V_{DD} - V_{T,n}$. With the voltage at the node V_X reaches $V_{X(max)}$ the pass transistor terms off since the gate to source voltage is now equal to its threshold voltage. So we see that the voltage at node. V_X cannot attain the full scale value of V_{DD} and is always attains a value of $V_{DD} - V_{T,n}$ for a logic 1 transfer.

If three pass transistors are cascaded (shown in figure below) and output of one acts as the input of the second one. If the initial input voltage is V_{DD}. Transistor M_1 is in saturation and the voltage at node 1 is $V_{DD} - V_{Tn1}$.

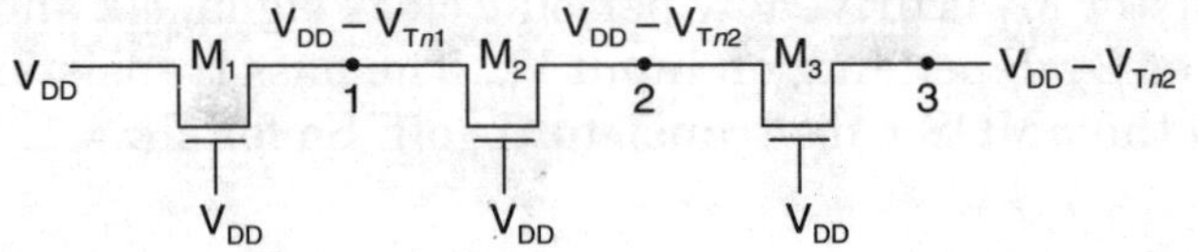

Assuming that the pass transistors used are identical the 2nd pairs transistors M_2 operates at saturation boundary and the voltage at node 2 is again one threshold voltage less than the supply voltage (max. voltage V_{DD}) ; $V_{2\,(max)} = V_{DD} - V_{Tn2}$ and so on.

If the threshold voltage of all the transistors are equal. It is noticed that, at the output will become one threshold voltage lower than V_{DD} and is independent of the number of pass transistors in the chain.

(*b*) **Logic '0' transfer:** If we assume that the voltage at node V_X $(t = 0) = V_{max} = V_{DD} - V_{T,n}$ initially then the capacitor is charged.

If a logic '0' is applied at the input *i.e.*, $V_{in} = 0V$ and the clock signal at the gate of the pass transistor is active then the transistor starts conducting. The drain current now flows from the internal terminal V_X to the input terminal. This means that the node V_X is now acting as drain and the input terminal acts as source (as shown in figure below).

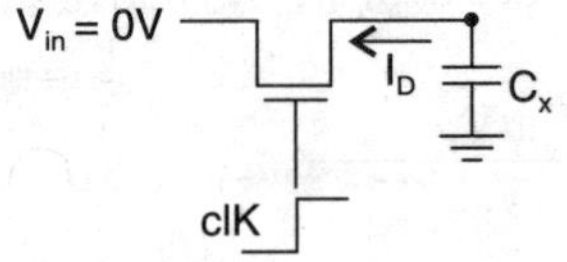

In this situation $V_{as} = V_{DD}$ and $V_{DS} = V_{max}$ (voltage at source = 0).
Since $V_{DS} < V_{as} - V_{T,n}$, the pass transistor M_1 is in linear region.
The pass transistor now discharges the capacitor C_X as,

$$-C_X \frac{dV_X}{dt} = \frac{K_n}{2}\left(2\left(V_{DD} - V_{Tn}\right)V_X - V_X^2\right) \qquad ...(iv)$$

$$dt = -\frac{2C_X}{K_n} \cdot \frac{dV_X}{2(V_{DD} - V_{Tn})V_X - V_X^2} \qquad ...(v)$$

Integrating equation (*v*), we get

$$\int_0^t dt = \frac{-2C_X}{K_n} \int_{V_{DD}-V_{Tn}}^{V_X} \left[\frac{\dfrac{1}{2(V_{DD}-V_{Tn})}}{2(V_{DD}-V_{Tn}) - V_X} + \frac{\dfrac{1}{2(V_{DD}-V_{Tn})}}{V_X}\right] dV_X$$

or,

$$t = \frac{C_X}{K_n\,(V_{DD} - V_{Tn})} \ln\left[\frac{2\,(V_{DD} - V_{Tn}) - V_X}{V_X}\right] \qquad ...(vi)$$

Equation (vi) can be written in terms of V_X and it can be shown that the voltage at node V_X drops from its logic high level of V_{max} to 0V.

Design Style:

A multistage synchronous circuit interconnected through nMOS pass transistor is shown in figure below:

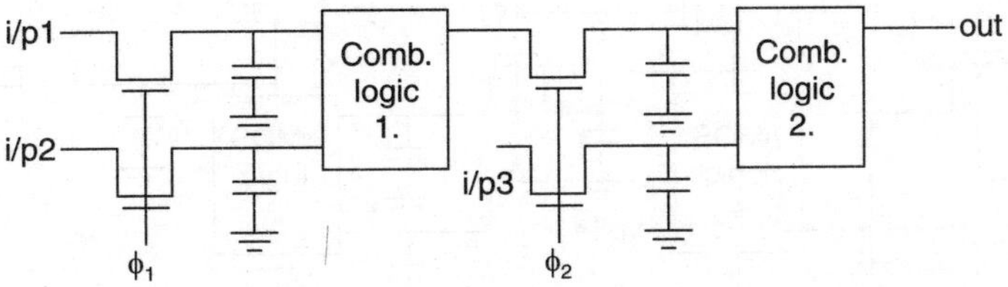

All the inputs of each combinational logic block are driven by a single clock signal. For the dynamic circuits, there have capacitances connected after every nMOS pass transistors and temporary charge storage takes place in these capacitances. Two non-overlapping clock signals ϕ_1 and ϕ_2 drive the pass transistors that is at a given time only one clock is active and is driving only one set of input pass transistors.

This can be explained in details as —

(*a*) When ϕ_1 is active stage 1 is active only and the inputs i/p1 and i/p2 are applied to the combinational logic 1 and it produces certain outputs. But since ϕ_2 is inactive during that time, the pass transistors driven by ϕ_2 are in cut-off state and retain their previously set logic levels.

(*b*) During the next phase when ϕ_2 is active the input levels of stage 2 will be applied to the combinational logic 2 through the pass transistors. During this time the input capacitances of stage 1 retain their logic levels.

Example:

$$Y = \overline{\overline{(A+B)C}\,.\,D}$$

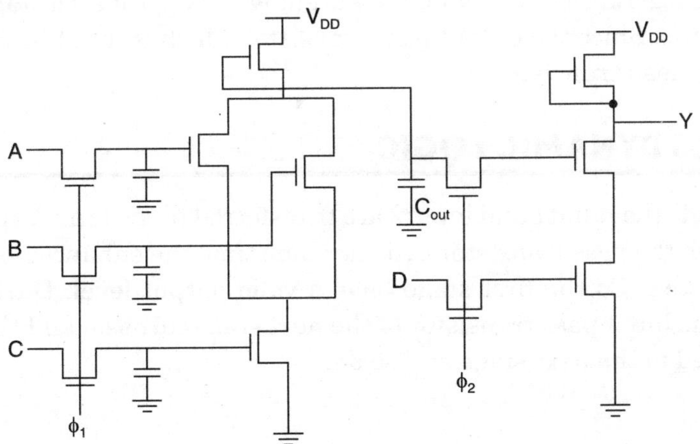

5.5 RATIOED DYNAMIC LOGIC

The clock signal can be applied to the gate of the load transistor instead of bearing the load transistors with a constant gate voltage as in depletion load *n*MOS synchronous logic.

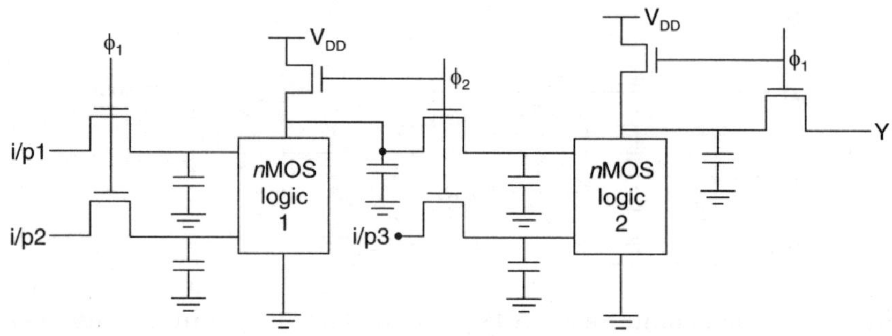

For the above reason, the power dissipation and chip area is significantly reduced in this approach.

When ϕ_1 is 'High', the input voltage level input1 and input2 are transferred to the first stage through the pass transistors. During this time, the enhancement type*n*MOS load transistor of the first stage *n*MOS logic 1 is not active yet.

During the next phase when ϕ_2 is 'High', the load transistor is turned 'on' and the input logic level is still preserved in the input capacitances, then the output of the first *n*MOS logic stage attains its valid logic level.

At the same time, the input pass transistors of the 2nd stage is also turned on which transfer this recently obtained output level to be transferred to the input capacitance of stage 2.

When ϕ_1 is 'High' again, the valid output level of stage 2 is determined and passed to the output node Y.

At this point new values of the inputs input1 and input2 are also accepted by the stage 1 and this process will continue.

In this type of logic design, the valid low output level (V_{OL}) of each stage is determined by the driver to load ratio. Since the output pass transistor which is input to the next stage turns on in phase with the load transistor.

5.6 RATIOLESS DYNAMIC LOGIC

When ϕ is 'high' the input1 and input2 are transferred to the input capacitances of*n*MOS logic stage 1 through the pass transistors. At the same time the enhancement type *n*MOS load transistor is also active. So the first stage have a valid output level. During the next phase when ϕ_2 is active, the input pass transistor of the next stage turns on and the logic levels at its input are transferred to the next stage and so on.

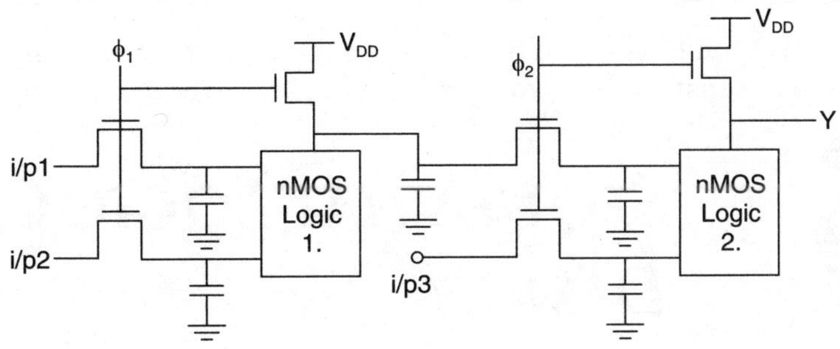

Two cases arise:

(*a*) If the output level across the output capacitor of stage 1 is logic high at the end of active ϕ_1, this voltage level is transferred to the input capacitance of the stage 2 during active ϕ_2 by charge sharing over the pass transistor.

But the high output level of first stage is subject to one threshold voltage drop $(V_{DD} - V_{Tn})$. To correctly transfer a logic high after charge sharing between the output and input capacitor between two stages the ratio of the capacitors (C_{output}/C_{input}) must be made large.

(*b*) If the output level of the first stage is low then the output capacitor (after stage 1) is completely discharged and V_{OL} = 0V when ϕ_1 turns off.

When ϕ_2 is active this logic low level is transferred to the next stage through the pass transistor connected between the stage 1 and stage 2.

Since a valid logic low level of 0V can be achieved irrespective of the driver to load ratio, this circuit arrangement is called ratioless logic.

Chapter 6

SEQUENTIAL LOGIC CIRCUIT DESIGN

6.1 INTRODUCTION

If we neglect the propagation delay time, the output of the combinational logic circuit, at any given time point are directly determined as Boolean function of the input variables applied at that time. Thus, the combinational circuits lack of the capability of storing any previous events, or displaying an output behaviour which is dependent upon the previously applied inputs. The sequential circuit gives the output, which is determined by the current inputs as well as the previously applied input variables.

Bistable circuits have, as their name implies, two stable stages or operation modes, each of which can be attained under certain input and output condition.

Monostable circuits, have only one stable operating point (state).

All basic Latch and flip-flop circuits, registers and memory elements used in digital systems fall into this category.

6.2 BEHAVIOUR OF BI-STABLE ELEMENTS

The basics of a bi-stable elements is that, it has two identical cross-coupled inverter circuits, as shown in figure given below:

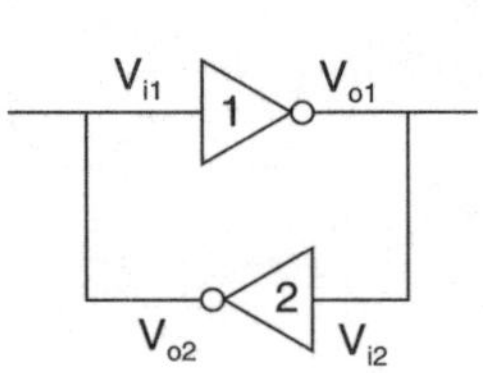

(a)

Here the output voltage of inverter (1) is equal to the input voltage of inverter (2).

$\therefore$ $V_{o1} = V_{i2}$ and the output voltage of inverter (2) is equal to the input voltage of inverter (1).

$\therefore$ $V_{o2} = V_{i1}$.

The voltage transfer characteristic curve of a bi-stable element is shown in figure as follows:

It can be seen that the two voltage transfer characteristics intersect at three points (two stable and one unstable points).

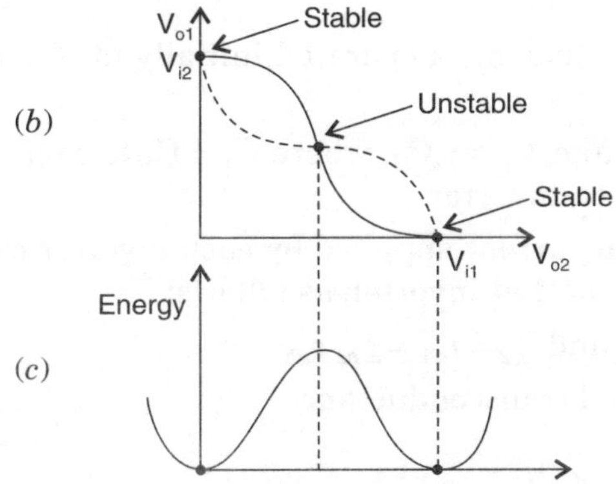

Qualitative view of the potential energy levels corresponding to the three operating points.

If the circuit is initially operating at one of these two stable points, it will preserve this state unless it is forced externally to change its operating point.

The gain of each inverter circuit, *i.e.,* the slope of the respective VTC, is smaller than unity at the two stable operating points. So, if we want to change the state by moving the operating point from one stable point to another, then a sufficiently large external voltage perturbation must be applied so that the voltage gain of the inverter loop becomes larger than unity.

On the other hand, the voltage gains of the both inverters are larger than unity at the third operating point. Consequently, even if the circuit is biased at this point initially, a small voltage perturbation at the input of any of the inverters will be amplified, causing the operating point to move to one of the stable operating points. This leads to the conclusion that the third operating point is unstable.

The circuit has two stable operating points, hence it is called bi-stable.

The circuit diagram of a CMOS bistable element (*a*) and the expected time domain behaviour of the output voltages, if the circuit is initially set at its unstable operating point (*b*) are, shown in figure given below:

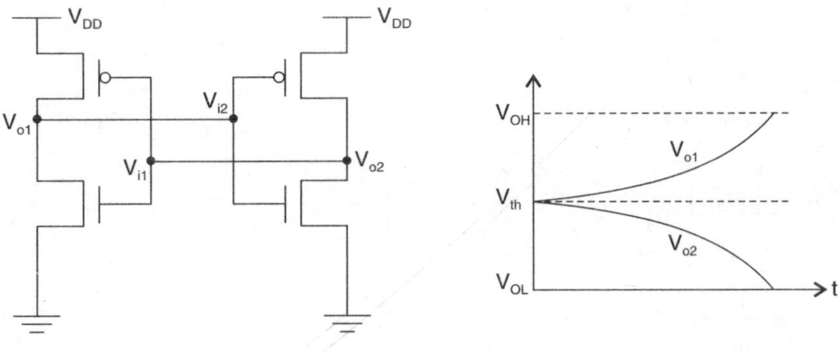

(*a*) (*b*)

Mathematical Analysis:

Consider the bistable element is operated initially at $V_{o1} = V_{o2} = V_{th}$, at the unstable operating point.

For the analysis, we take, $C_g >> C_d$, where C_g = Gate capacitance of each inverter, and C_d = Drain capacitance of each inverter.

The small-signal drain current supplied by each inverter can be expressed, in terms of the small signal gate voltage of that inverter, as follows:

So, $i_{g1} = i_{d2} = g_m\, v_{g2}$ and $i_{g2} = i_{d1} = g_m\, v_{g1}$ ...(*i*)

where g_m = small-signal transconductance.

$$v_{g1} = \frac{q_1}{c_g} \text{ and } vg_2 = \frac{q_2}{c_g} \quad ...(ii)$$

$$\therefore \quad \left.\begin{aligned} i_{g_1} &= c_g \frac{dv_{g1}}{dt} \\ i_{g_2} &= c_g \frac{dv_{g2}}{dt} \end{aligned}\right\} \quad ...(iii)$$

Combining equation (*i*) and (*iii*), we get

$$g_m\, v_{g2} = c_g \frac{dv_{g1}}{dt} \quad ...(iv)$$

$$g_m\, v_{g1} = c_g \frac{dv_{g2}}{dt} \quad ...(v)$$

Expressing the gate voltages in terms of the gate charges, there two differential equations can also be written as,

$$\frac{g_m}{c_g}\, q_2 = \frac{dq_1}{dt} \quad ...(vi)$$

$$\frac{g_m}{c_g}\, q_1 = \frac{dq_2}{dt} \quad ...(vii)$$

$$\Rightarrow \quad \frac{g_m}{c_g}\, q_1 = \frac{c_g}{g_m} \frac{d^2 q_1}{dt^2}$$

$$\Rightarrow \quad \frac{d^2 q_1}{dt^2} = \left(\frac{g_m}{c_g}\right)^2 q_1 \quad ...(viii)$$

$$= \frac{1}{\tau_0^2}\, q_1 \quad ...(ix)$$

where $\tau_0 = \dfrac{c_g}{g_m}$; ....

The time domain solution of (ix) for q_y, we get,

$$q_1(t) = \frac{q_1(0) - \tau_0 q_1'(0)}{2} e^{-t/\tau_0} + \frac{q_1(0) - \tau_0 q_1'(0)}{2} e^{t/\tau_0} \qquad ...(x)$$

where the initial condition is given by,

$$q_1(0) = c_g\, v_{g1}(0).$$

Here, $v_{g1} = v_{02}$ and $v_{g2} = v_{01}$.

So, $$V_{02}(t) = \frac{1}{2}\{V_{02}(0) - \tau_0 V_{02}'(0)\} e^{-t/\tau_0} + \frac{1}{2}\{V_{02}(0) + \tau_0 V_{02}'(0)\} e^{t/\tau_0}$$

$$V_{01}(t) = \frac{1}{2}\{V_{01}(0) - \tau_0 V_{01}'(0)\} e^{-t/\tau_0} + \frac{1}{2}\{V_{01}(0) + \tau_0 V_{01}'(0)\} e^{t/\tau_0}$$

For large values of t, we can write,

$$V_{01}(t) = \frac{1}{2}\{V_{01}(0) + \tau_0 V_{01}'(0)\} e^{t/\tau_0} \text{ and } V_{02}(t) = \frac{1}{2}\{V_{02}(0) + \tau_0 V_{02}'(0)\} e^{t/\tau_0}$$

The magnitude of both output voltages increases exponentially with time.

Depending on the polarity of the initial small perturbations $V_{O1}(0)$ and $dv_{O2}(0)$, the output voltages of both inverters will diverge from their initial value of v_{th} to either V_{OL} or V_{OH}.

The two operating points ($v_{o1} = V_{OL}$, $v_{o2} = V_{OH}$) and $v_{o1} = V_{OH}$, $v_{o2} = V_{OL}$) can shown to be stable and the point ($v_{o1} = V_{th}$, $v_{o2} = V_{th}$) is unstable.

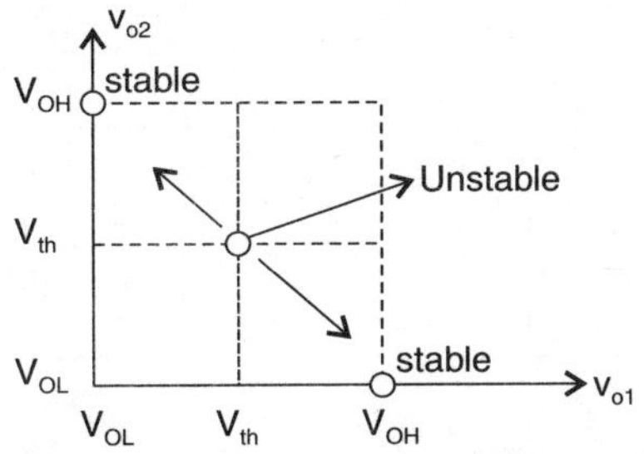

Phase-plane representation of the bistable circuit behaviour.

6.3 S-R LATCH CIRCUIT

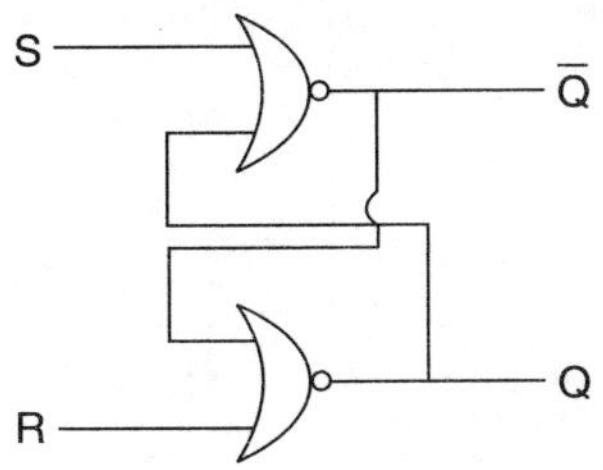

Truth Table:

SR	Q_{n+1}	Q_{n+1}	*Operation*
0 0	Q_n	$\overline{Q_n}$	hold
0 1	0	1	reset
1 0	1	0	set
1 1	?	?	not allowed

The CMOS SR Latch circuit based on NOR2 Gate:

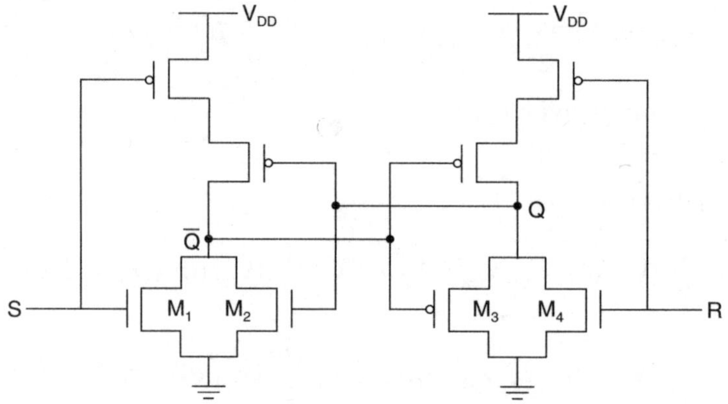

The operation of the circuit is given in table shown below:

S	*R*	Q_{n+1}	$\overline{Q_{n+1}}$	*Operation*
V_{OH}	V_{OL}	V_{OH}	V_{OL}	M_1 and M_2 on, M_3 and M_4 off.
V_{OL}	V_{OH}	V_{OL}	V_{OH}	M_1 and M_2 off, M_3 and M_4 on.
V_{OL}	V_{OL}	V_{OH}	V_{OL}	M_1 and M_4 off, M_2 on or M_1 and M_4 off, M_3 on.

Depletion load *n*MOS SR Latch circuit based on NOR2 Gate:

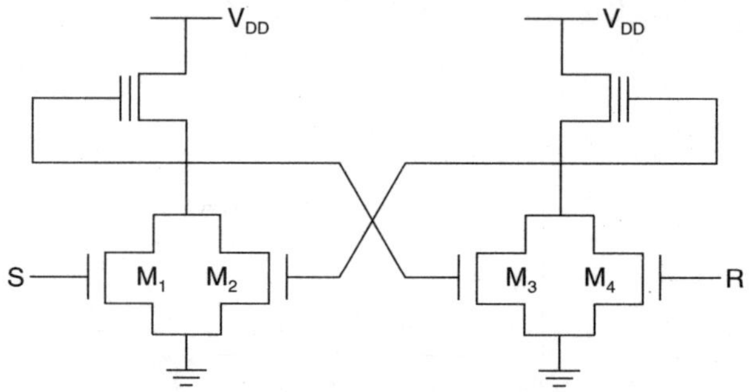

CMOS SR Latch circuit based on NAND Gate:

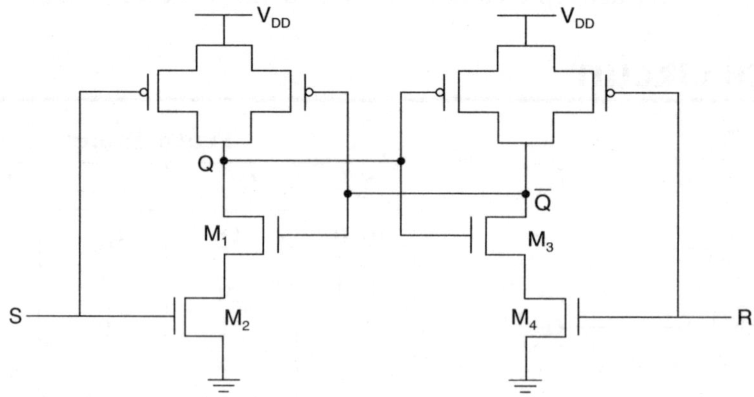

Truth Table:

S	R	Q_{n+1}	$\overline{Q_{n+1}}$	*Operation*
0	0	?	?	not allowed
0	1	1	0	set
1	0	0	1	reset
1	1	on	on	hold

Operation:

S	R	Q_{n+1}	$\overline{Q_{n+1}}$	*Operation*
V_{OL}	V_{OH}	V_{OH}	V_{OL}	M_1 and M_2 off, M_3 and M_4 on.
V_{OH}	V_{OL}	V_{OL}	V_{OH}	M_1 and M_2 on, M_3 and M_4 off.
V_{OH}	V_{OH}	V_{OL}	V_{OH}	M_2 and M_4 on, M_1 on or,
V_{OH}	V_{OH}	V_{OH}	V_{OL}	M_2 and M_4 on, M_3 on.

6.4 CLOCKED S-R LATCH

Truth Table:

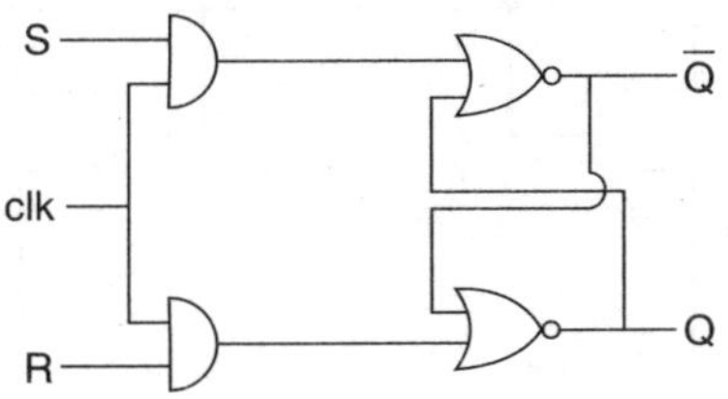

CMOS Circuit:

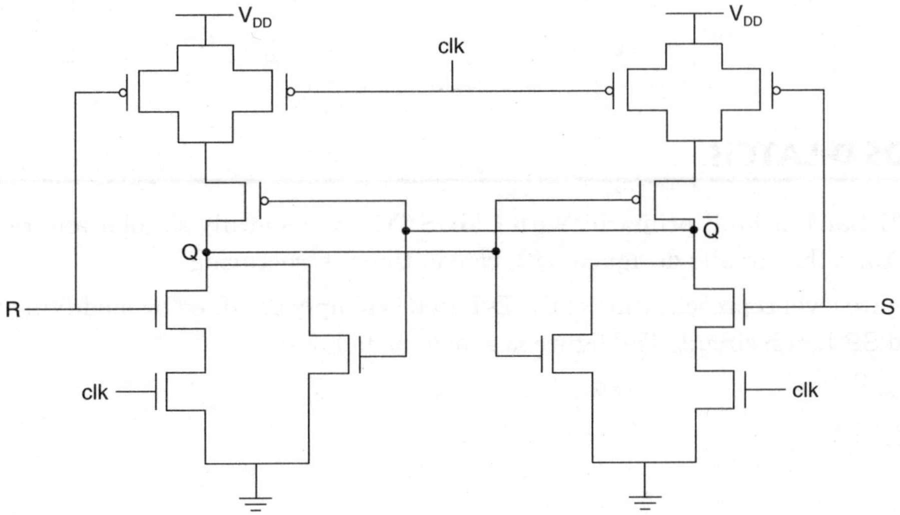

Clocked J-K Latch:

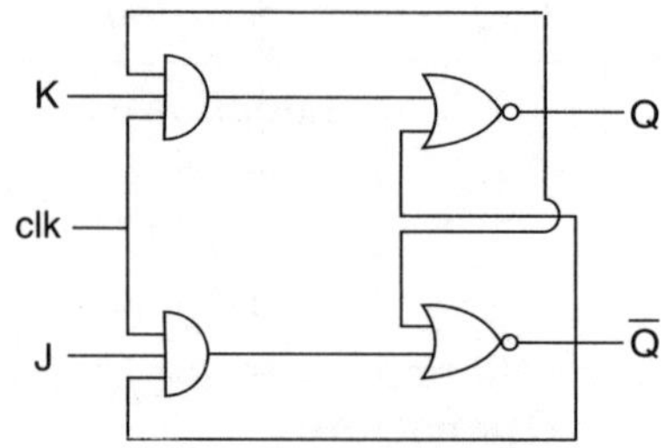

CMOS J-K Latch:

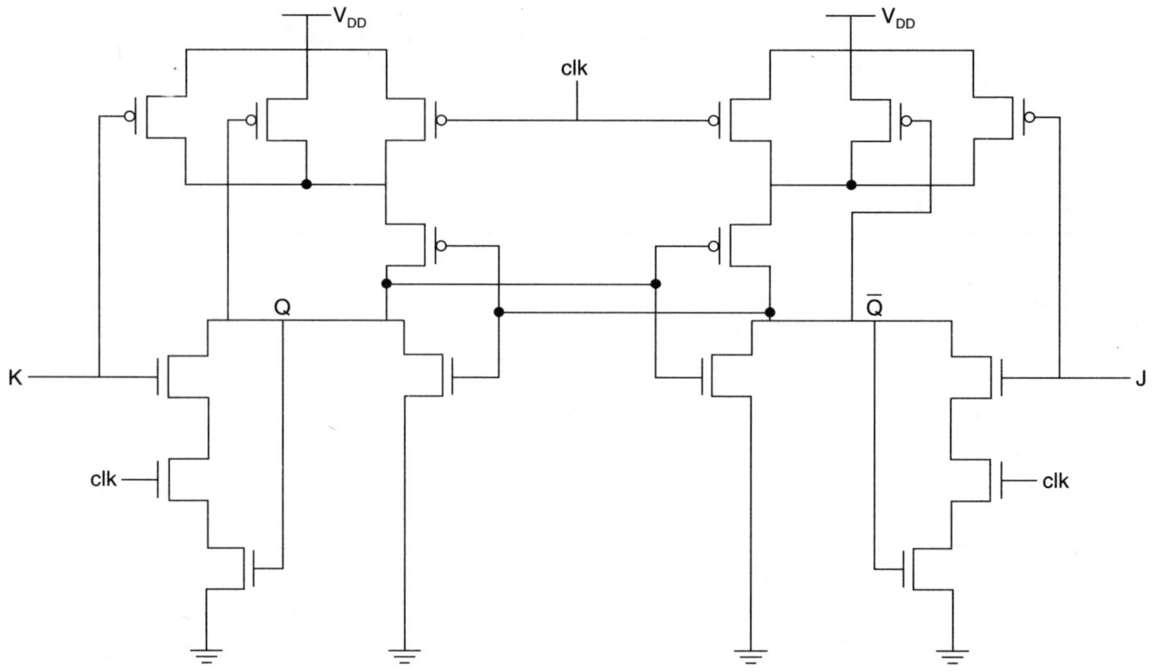

6.5 CMOS D-LATCH

The D-Latch is built primarily with CMOS TG are generally simpler and require fewer transistors than the circuits designed with conventional structuring.

The gate level representation of the D-Latch is simply obtained by modifying the clocked NOR–based SR Latch circuit. The figure is shown as follows:

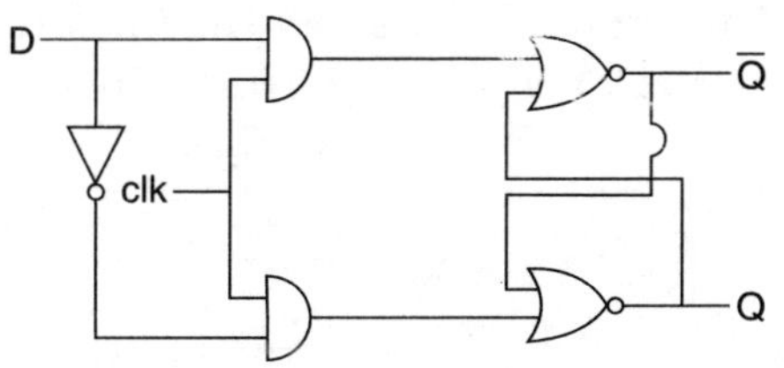

The output Q assumes the value of the input D where clk = '1'. When clk = '0', the output will simply preserve its state. Thus, the clk input acts as an enable signal which allows data to be accepted into the D-Latch.

6.6 THE TG BASED D-LATCH

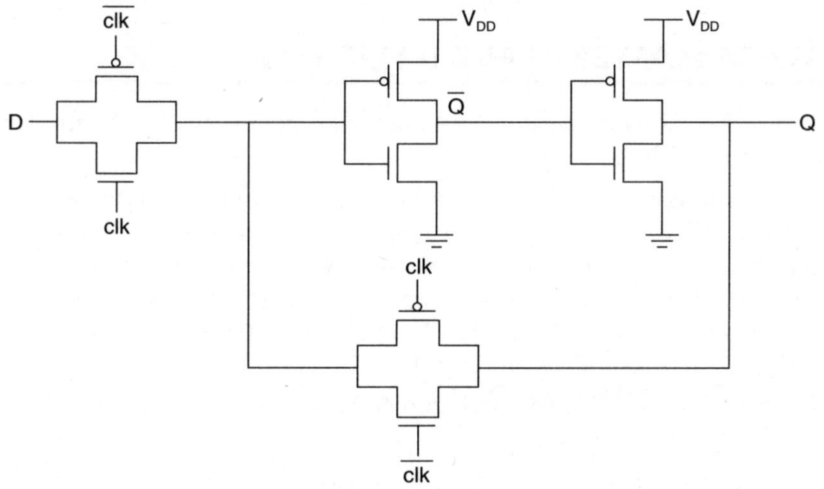

The TG at the input is activated by the clk signal, whereas the TG in the inverter loop is activated by the inverse of the clk signal ; $\overline{clk}$.

Thus the input signal is accepted (latched) into the circuit when the clock is high, and this information is preserved as the state of the inverter loop when the clk is low.

The timing diagram is shown in figure given below:

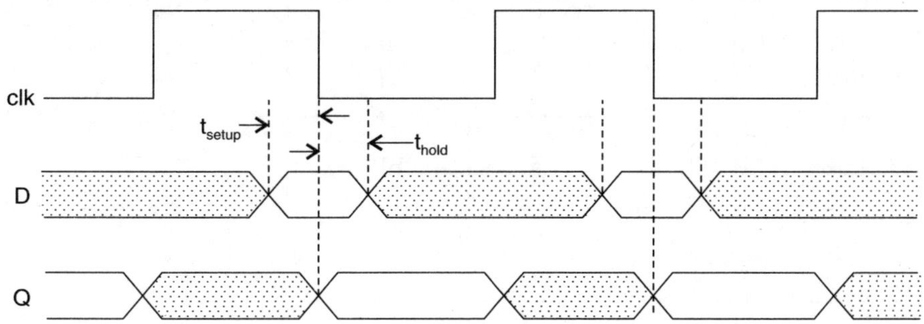

The valid D input must be stable for a short time before (setup time, t_{setup}) and after (holding time, t_{hold}) the negative clock transition, during which the input switch opens and the loop switch closes.

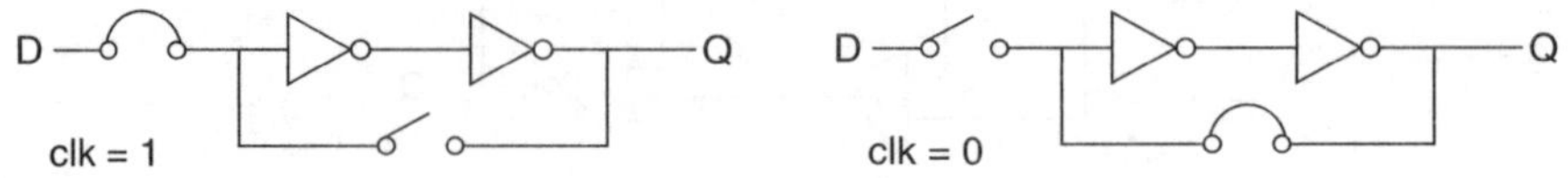

Once the inverter loop is completed by closing the loop switch, the output will preserve its valid level.

Note. In D-Latch design, the requirements for setup time and hold time should be met carefully. Any violation of such specifications can cause metastability problems, which lead to very extreme chaotic transient behaviour and can result in an unpredictable state after the transitional period.

6.7 TWO STAGE MASTER-SLAVE D FLIP-FLOP

Operation. When the clk is *high*, the master stage follows the D input while the slave stage holds the previous value.

When the clk changes from logic '1' to logic '0', the master latch ceases to sample the input and stores the D value at the time of the clock transition. At the same time, the slave latch becomes transparent, passing the stored master value Q_m to the output of the slave stage, Q_s. The input cannot affect the output because the master stage is disconnected from the D input. When the clock changes again from logic '0' to '1' the slave latch locks in the master latch ouput and the master stage starts sampling the input again.

Thus, the circuit is a negative edge-triggered D flip-flop by virtue of the fact that it samples the input at the falling edge of the clk pulse.

The master-slave D.F.F. is shown in figure below:

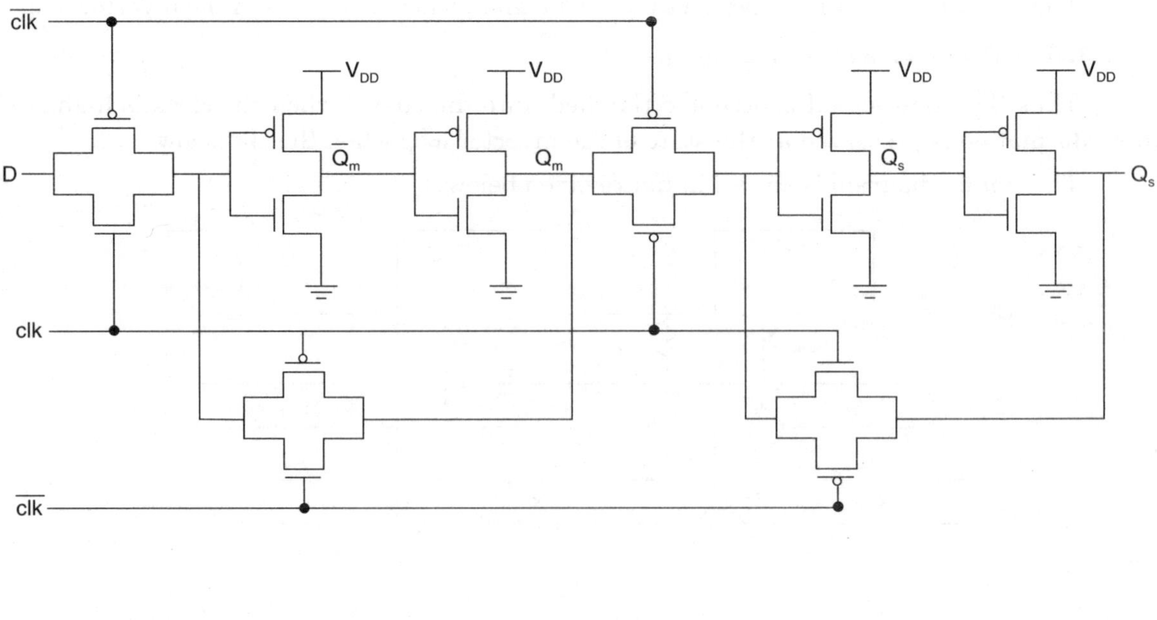

Chapter 7

ARITHMETIC SUBSYSTEM

7.1 CARRY LOOKAHEAD ADDER

The linear growth of adder carry delay with the size of the input word may be improved by calculating the carries to each stage in parallel.

The carry of the ith stage, C_i may be expressed as,

$$C_i = G_i + P_i \, . \, C_{i-1} \qquad ...(i)$$

where

$$G_i = A_i \, . \, B_i \qquad \text{generate signal} \quad ...(ii)$$

$$P_i = A_i \oplus B_i \qquad \text{propagate signal} \quad ...(iii)$$

Expanding this, we get,

$$C_i = G_i + P_i \, G_{i-1} + P_i \, P_{i-1} \, G_{i-2} + ... + P_i \, P_{i-1} ...... P_1 \, C_0 \qquad ...(iv)$$

The sum S_i is generated by

$$S_i = C_{i-1} \oplus A_i \oplus B_i \qquad ...(v)$$

if

$$P_i = A_i \oplus B_i$$

For four stages of lookahead, the appropriate terms are

$$C_1 = G_1 + P_1C_0$$

$$C_2 = G_2 + P_2G_1 + P_2P_1C_0$$

$$C_3 = G_3 + P_3G_2 + P_3P_2G_1 + P_3P_2P_1C_0$$

$$C_4 = G_4 + P_4G_3 + P_4P_3G_2 + P_4P_3P_2G_1 + P_4P_3P_2P_1C_0$$

$$= G_4 + P_4 \, [G_3 + P_3\{G_2 + P_2(G_1 + P_1C_0)\}] \qquad ...(vi)$$

A possible implementation of the carry gate for this kind of carry lookahead adder for 4 bits is shown in Fig. 7.1(a). Note that the gates have been partitioning to keep the number of inputs less than or equal to four. This is typical of the type of carry lookahead that would be used in a gate array or standard cell design.

Fig. 7.1(b) designed by the above expression as a domino CMOS gate. Carry $C_1 - C_3$ are generated similarly.

Note that the worst care delay path in this circuit, has six n-MOS in series.

A static version of the C_4 gate is shown in Fig. 7.1(c). The symbolic layout of a gate to implement this function is shown in Fig. 7.1(d).

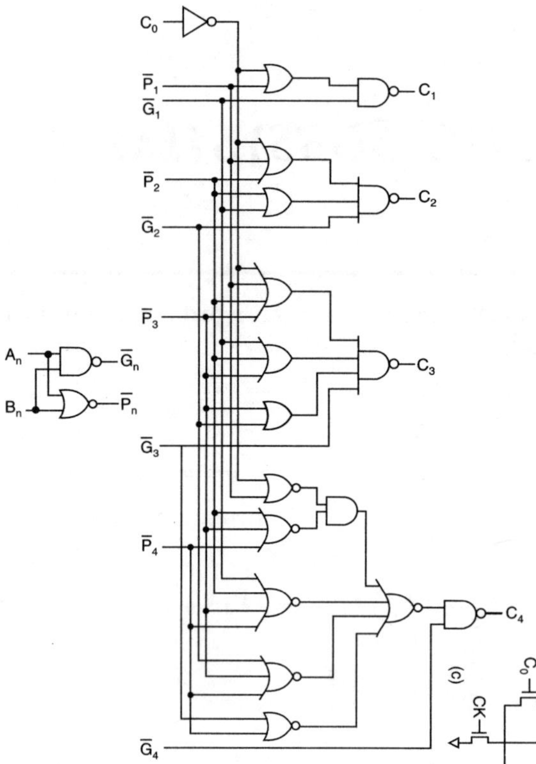

(a) Carry Lookahead Adder (Basic Circuit)

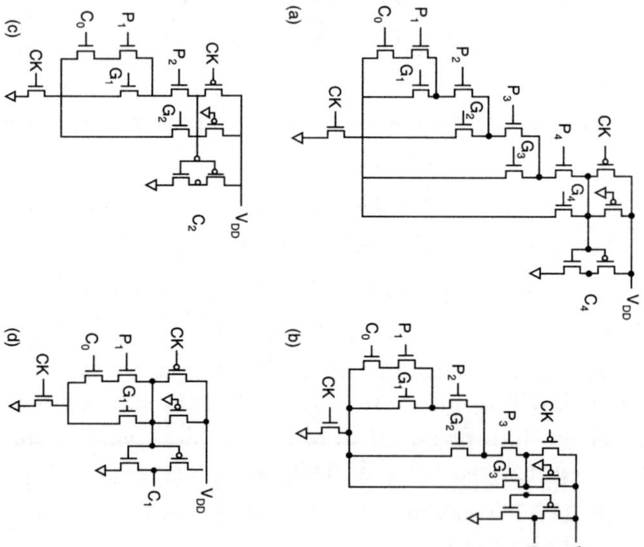

(b) Domino Carry Lookahead

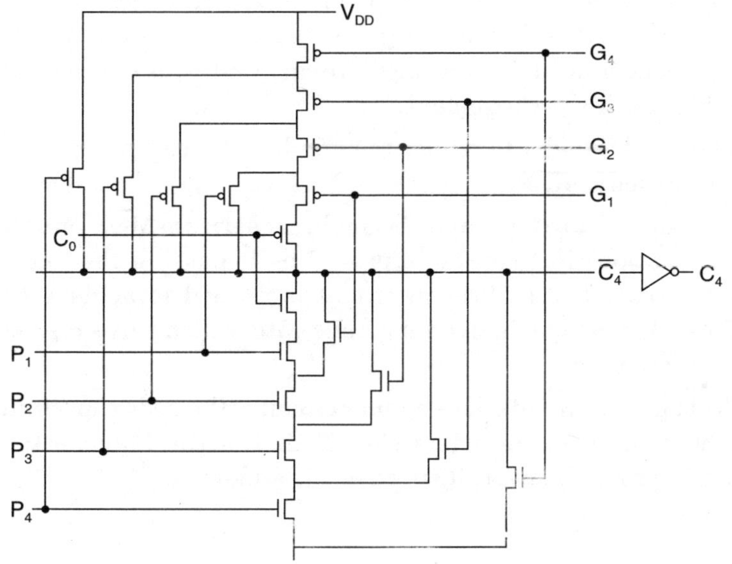

(c) Static Carry Lookahead

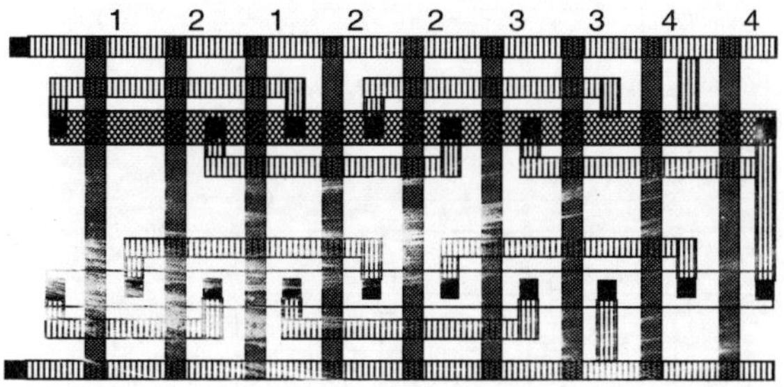

(d) Layout of a Static Carry Lookahead Gate

Fig. 7.1

7.2 MANCHESTER CARRY ADDER

The efficiency of the domino carry chain can be enhanced by precharging at appropriate points. The elemental circuit is shown in Fig. 7.2(a).

Operation:

When CLOCK is low, the output node is precharged by the p pull-up transistor. When CLOCK goes high, the n pull-down transistor turns on. If carry generate (AB) is true, then the output node discharges. If carry propagate (A + B) is true, then a previous carry may be coupled to the output node, conditionally discharging it. Note that in this circuit $\overline{\text{CARRY}}$ is actually propagated.

We can construct a 4-bit adder by cascading four such stages and constructing the circuitry to supply the appropriate signals. This is commonly called Manchester Carry Adder. Thus a 4 bit adder would be constructed as shown in Fig. 7.2(*b*). Notice the similarity to the domino carry circuit. However, we no longer need the intermediate carry gates, as the carry values are available in a distributed fashion in this arrangement.

We have chosen a 4-bit adder to reduce the number of series propagate transistors, which reduces the influence of body effect.

If all propagate signals are true, and C is high, six series *n*-MOS pull the output node low. This worst care propagation time can be improved by bypassing the four stages if all carry propagate signals are true. The additional circuit is needed to achieve this. This circuit is expressed in Fig. 7.2(*c*). It consist of a dynamic AND gate, which turns on a carry bypass signal if all carry propagates are true.

Note that, although the circuits have similar circuitry the node capacitance at intermediate nodes in the lookahead gate is approximately 1/2 that of the Manchester chain. Thus this arrangement should improve the overall speed of the adder.

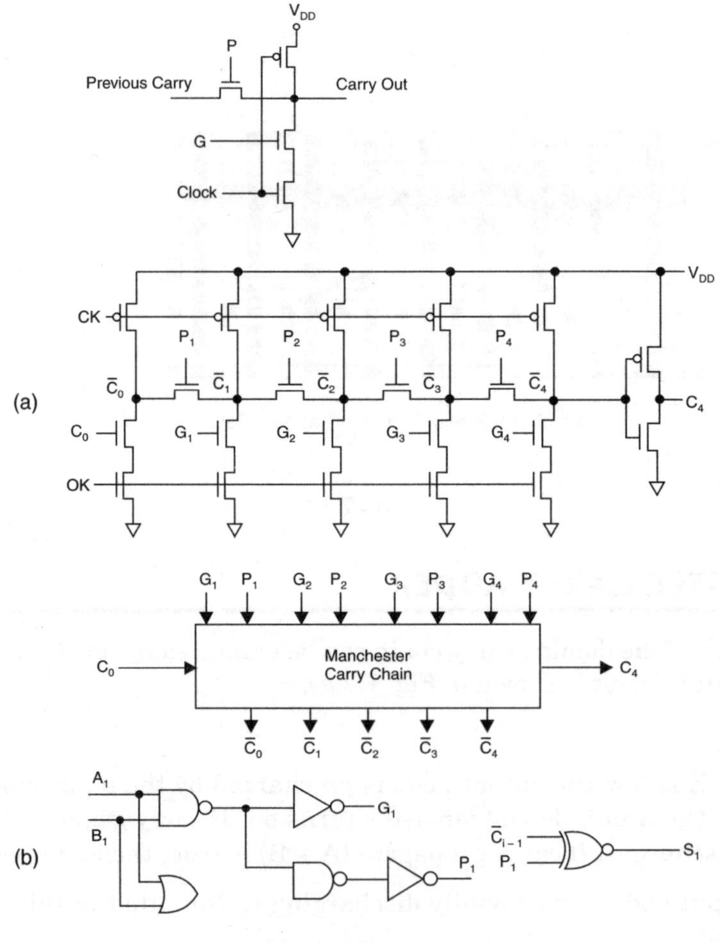

(a) and (b) Manchester Carry Chain

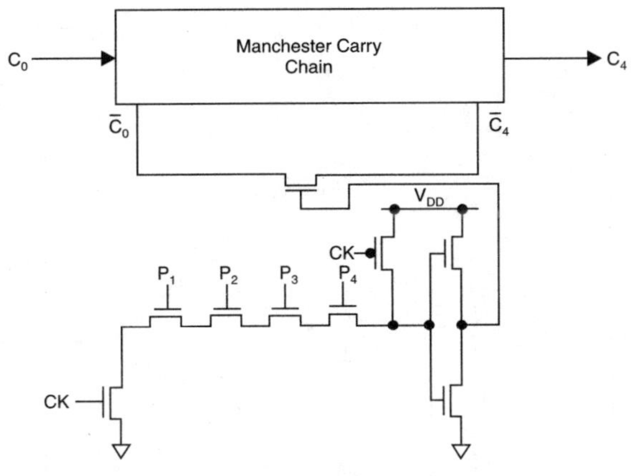

(c) Manchester Lookahead Circuitry

Fig. 7.2

7.3 STATIC RAM

A low power SRAM cell may be designed simply by using cross coupled CMOS inverters.

The data storage cell, *i.e.*, the 1 bit memory cell in static RAM arrays, consists of a simple latch circuit with two stable operating points. Depending on the preserved state of the two inverter latch circuit, the data being held in the memory cell will be interpreted either as a logic '0' or as a logic '1'.

To access (read or write) the data content in the memory cell *via* the bit line, we need atleast one switch which is controlled by the corresponding word line, *i.e.*, the row address selection signal.

Usually, two complementary access switches consisting of *n*MOS pass-transistors are implemented to connect the 1-bit SRAM cell to the complementary bit lines (columns).

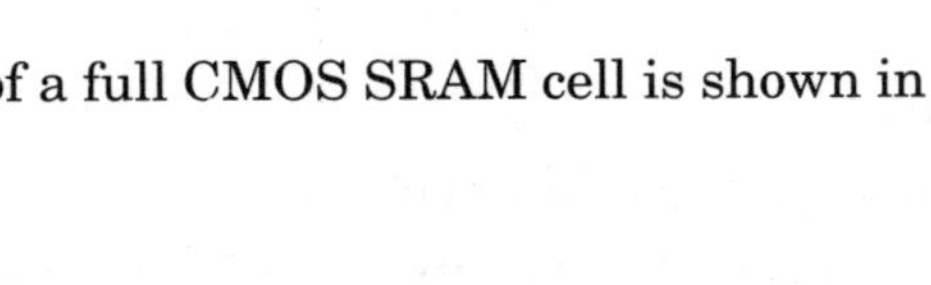

Fig. 7.3

Full CMOS SRAM Cell:

The circuit diagram of a full CMOS SRAM cell is shown in the figure given as follows :

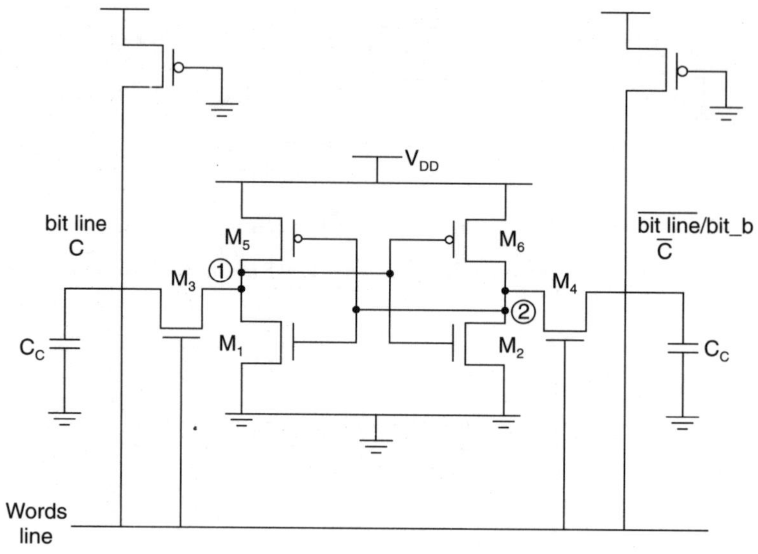

Fig. 7.4

Advantages:

(*i*) Lowest static power consumption due to the small leakage currents of the each inverter.

(*ii*) High noise margin.

Disadvantages:

(*i*) Larger cell size.

(*ii*) Tendency to exhibit "Latch up" phenomena.

Structure:

There are the *p*MOS column pull up transistors are attached on the complementary bit lines.

The memory cell consists of a simple CMOS Latch (two inverters connected back to back) and two complementary access transistors (M_3 and M_4).

The cell will preserve one of its two possible stable states, as long as the power supply is available.

The access transistors are turned on whenever a word line (row) is activated for read or write operation, connecting the cell to the complementary bet-line column.

Operation:

Data Read Operation:

Assuming that a logic '0' is stored in the cell.

Voltage levels in the SRAM cell at the beginning of read operation is shown in Fig. 7.5.

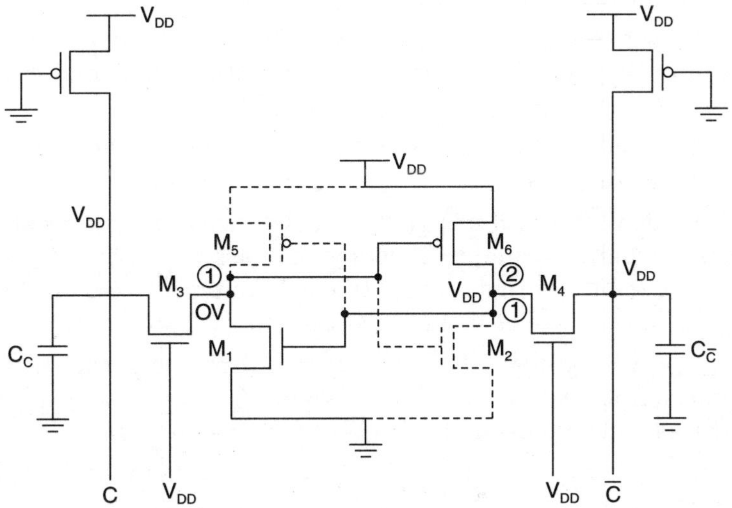

Fig. 7.5

Here, the transistors M_2 and M_5 are turned off, while the transistors M_1 and M_6 operate in the linear mode. Thus, the internal node voltages are $V_1 = 0$ and $V_2 = V_{DD}$ before the cell access (or pass) transistors M_3 and M_4 are turned on.

After the pass transistors M_3 and M_4 are turned on by the row selection circuitry, the voltage level of column $\bar{c}$ will not show any significant variation since no current will flow through M_4.

On the other half of the cell, however, M_3 and M_1 will conduct a nonzero current and the voltage level of column C will begin to drop slightly.

The column capacitance C_C is typically very large, therefore, the amount of decrease in the column voltage is limited to a few hundred mV during the read phase.

The key design issue for the data read operation is then to guarantee that the voltage V_1 does not exceed the threshold voltage of M_2. So that the transistor M_2 remains turned off during the read phase.

Write Operation:

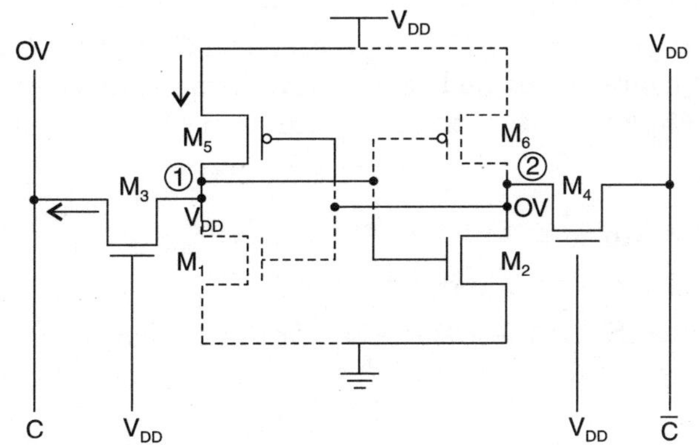

Fig. 7.6

Voltage levels in the SRAM cell at the beginning of the 'write' operation.

Now consider the write '0', operation, assuming that a logic '1' is stored in the SRAM cell initially. The transistors M_1 and M_6 are turned off, while the transistors M_2 and M_5 operate in the linear mode.

Thus the internal mode voltages are $V_1 = V_{DD}$ and $V_2 = 0V$ before the cell access (or pass) transistors M_3 and M_4 are turned on.

The column voltage V_e is forced to logic '0' level by the data write circuitry ; thus, we may assume that V_e is approximately equal to 0V. Once the pass transistors M_3 and M_4 are turned on, by the row selection circuitry, we expect that the node voltage V_2 remains below the threshold voltage of M_1, since M_2 and M_4 are designed according to $V_{1\,max} \leq V_{T2}$.

Consequently, the voltage level at node (2) would not be sufficient to turn on M_1. To change the stored information, *i.e.,* to force V_1 to 0V and V_2 to V_{DD}, the node voltage V_1 must be reduced below, the threshold voltage of M_2, so that M_2 turns off first.

This will guarantee that M_1 subsequently turns on, changing the stored information.

7.4 DYNAMIC RAM

Dynamic RAMs (DRAMs) store their contents as charge on a capacitor rather than in a feedback loop. Thus, the basic cell is substantially smaller than SRAM, but the cell must be periodically read and refreshed so that its contents do not leak away.

Advantage:

They offer an order of magnitude greater density (bits/cm^2) than high performance SRAM.

Disadvantage:

They have much higher latency.

7.4.1 One Transistor DRAM

A 1-transistor (1T) dynamic RAM cell consists of a transistor and a capacitor. Like SRAM, the cell is accessed by asserting the wordline to connect the capacitor to the bit line.

On a read, the bitline is first precharged to $V_{DD}/2$. When the wordline rises, the capacitor shares its charge with the bitline causing a voltage change ΔV. The read disturbs the cell contents at x, so the cell must be rewritten after each read.

On a write, the bitline is driven high or low and the voltage is forced onto the capacitor. Some DRAMs drive the wordline to $V_{PDP} = V_{DD} + V_t$ to avoid a degraded level when writing a '1'. The DRAM capacitor C_{cell} must be as physically small as possible to achieve good density.

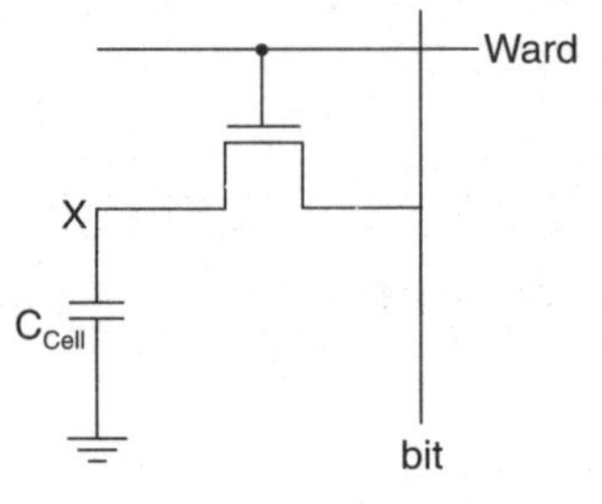

Fig. 7.7

7.4.2 3T-DRAM

The circuit diagram of a typical 3T-DRAM cell with column pull-up (precharge) transistors and the column read/write circuitry. Here the binary information is stored in the from of charge in the parasitic node capacitance C_1.

The storage transistor M_2 is turned on/off depending on the charge stored in C_1 and the pass transistors M_1 and M_3 act as access switches for data read and write operations. The cell has two separate bit lines for "data read" and "data write", and two separate word lines to control the access transistors.

The operation of the 3T-DRAM cell and its peripheral circuitry is based on a two-phase non-overlapping clock scheme. The precharge events are driven by ϕ_1, whereas the "read" and "write" events are driven by ϕ_2.

Every "data read" and "data write" operation is preceded by a precharge cycle, which is initiated with the precharge signal PC going high.

During the precharge cycle, the column pull-up transistors are activated, and the corresponding column capacitances C_2 and C_3 are charged upto logic-high.

The all "data read" and "data write" operations are performed during the active ϕ_2 phase, when PC is low.

Write 1 :

(*i*) $\overline{\text{DATA}}$ is at logic low level.

(*ii*) "data write" transistor MD is turned off.

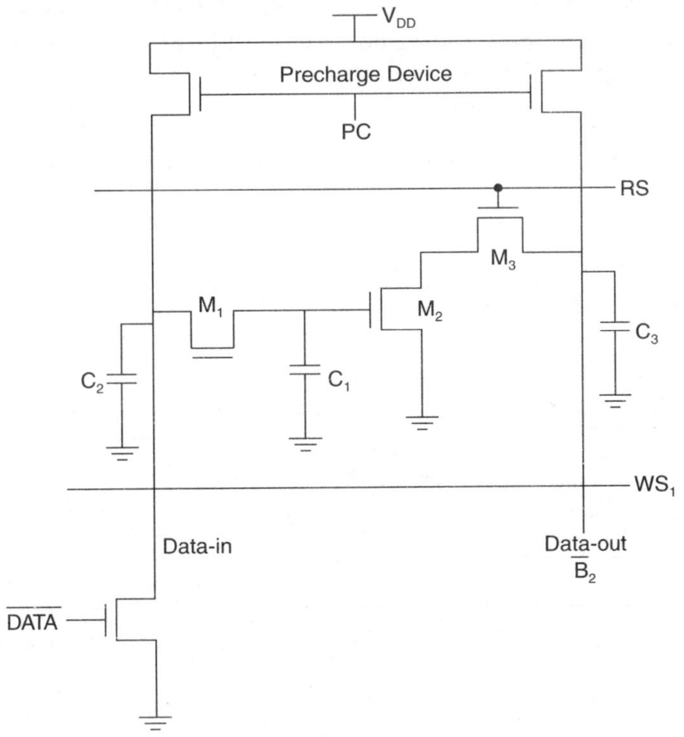

Fig. 7.8

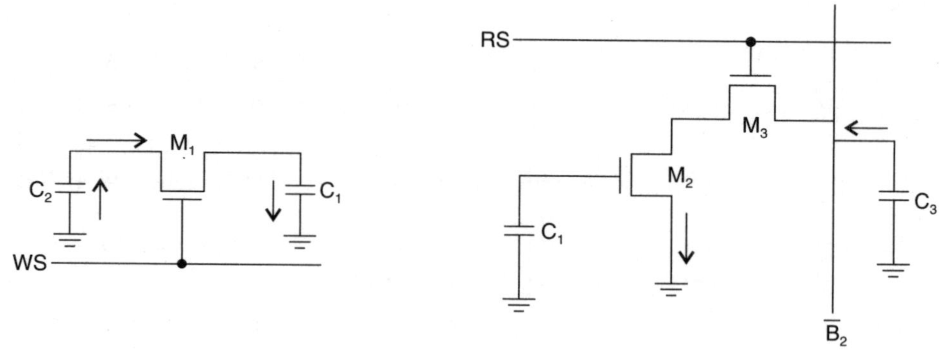

(*a*)

Charge sharing between C_2 and C_1 during the write '1' sequence.

(*b*)

The column capacitance C_3 is discharged through the transistors M_2 and M_3 during the read '1' operation.

(*iii*) Voltage level on column D_{in} remains high.

(*iv*) "Write select" signal WS is pulled high during the active phase of ϕ_2.

(*v*) The "write access" transistor is turned on.

(*vi*) The charge on C_2 is now shared with C_1.

(*vii*) Since the capacitance C_2 is very large compared to C_1, the storage node capacitance C_1 attains approximately the same logic-high level as the column capacitance C_2 at the end of the charge-sharing process.

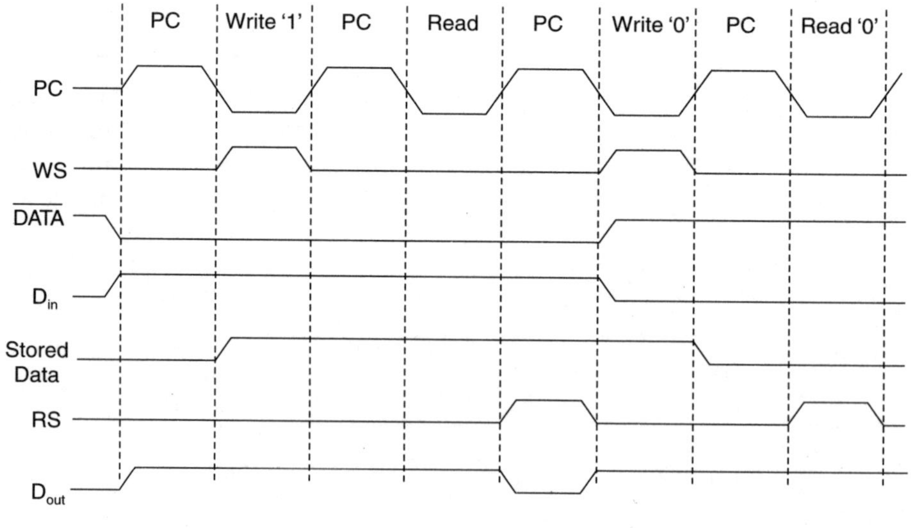

(*c*)

Fig. 7.9

Read 1:

(*i*) After the write '1' operation is completed, the write access transistor M_1 is turned off.

(*ii*) With the storage capacitance C_1 charged upto a logic high level, transistor M_2 is now conducting.

(*iii*) 'Read select' signal RS must be pulled high during the active phase of ϕ_2 following a precharge cycle.

(*iv*) Read access transistor M_3 turns ON, M_3 and M_2 create a conducting path between the data read column capacitance C_3 and ground.

(*v*) The capacitance C_3 discharges through M_2 and M_3 and the falling column voltage is interpreted by the 'data read' circuitry as a stored logic '1'.

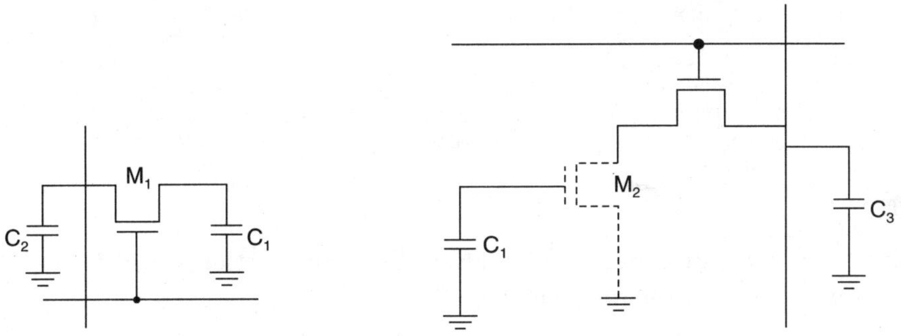

(*a*)

C_2 and C_1 both are discharged by M_1 at the write '0' sequence.

(*b*)

Column capacitance C_3 does not discharged at the read '0' sequence.

Fig. 7.10

7.5 READ ONLY MEMORY

It is a permanent or semi-permanent memory. In permanent ROM, the data is permanently stored and cannot be changed. It can only be read from the memory. There cannot be a write operation because the specified data is programmed into the device by the manufacturer or user. In semi-permanent ROM also there is no write operation, but the data can be altered by special method.

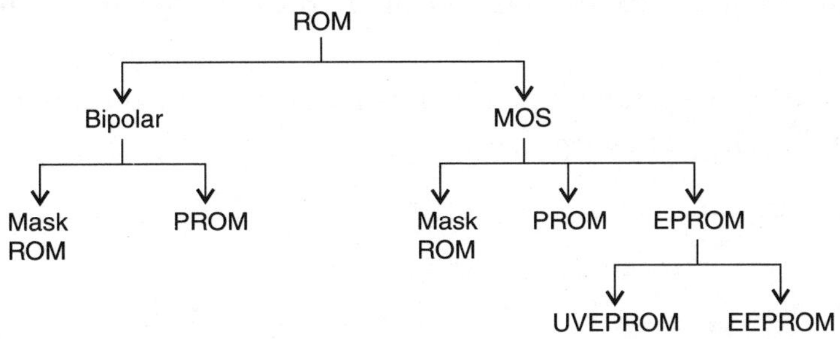

Fig. 7.11

Bipolar ROM:

It is made by BJT.

When a connection is present from a ROW line to base of transistor we get logical '1'. This is because of the reason that all transistors with base connection to the High Row line turn on and connect High (*i.e.,* 1) to the associated column line.

When there is no base connection at Row/Column junction, the column line remains Low (*i.e.,* 0) when the Row is addressed.

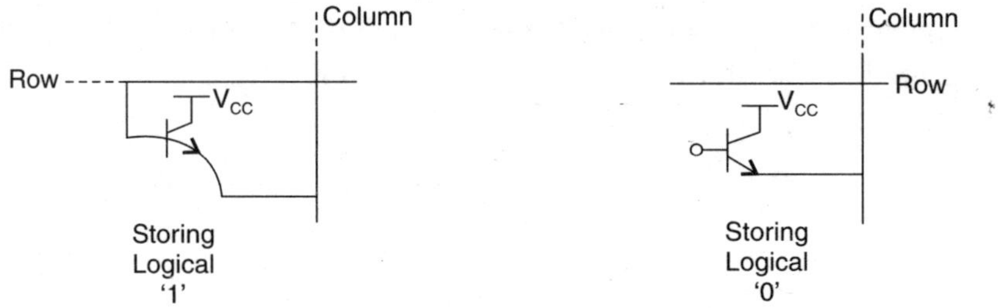

MOS ROM:

The presence of a gate connection at a Row-Column junction stores logical '1' permanently and absence of gate connection stores logical '0'.

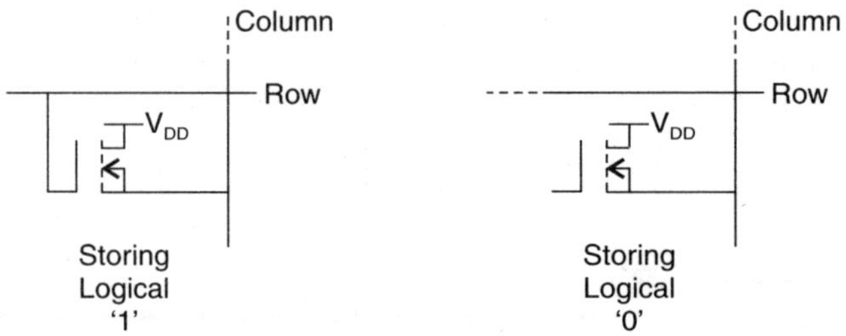

A 16 × 8 bit ROM is shown in figure below. Here Row line = 16 and Column line = 8.

The address input and decoder is also shown. At each Row-Column junction a transistor is connected. The shaded squares represent '0's and blank squares represent 1s. When a binary address code is applied to the address input, the corresponding Row line goes High.

When the transistor at a junction is storing '1', the corresponding column line also goes High.

At a cell storing '0', the column line remains Low due to the terminating resistance. The column lines forms the 8 bit data output.

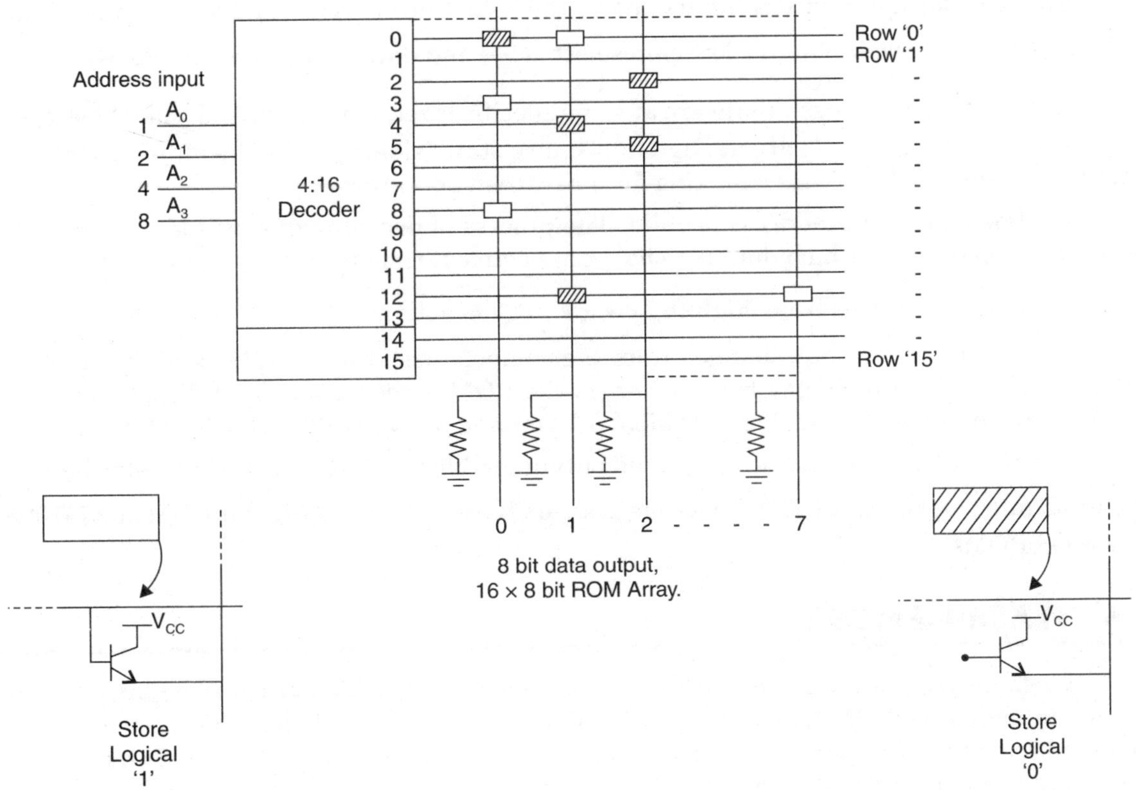

Fig. 7.12

ROM Access Time:

ROM access time is the time delay between the application of a valid address code on the inputs and the appearance of data output during a read operation. It is a measure of the operating speed of ROM. The waveform is shown given below.

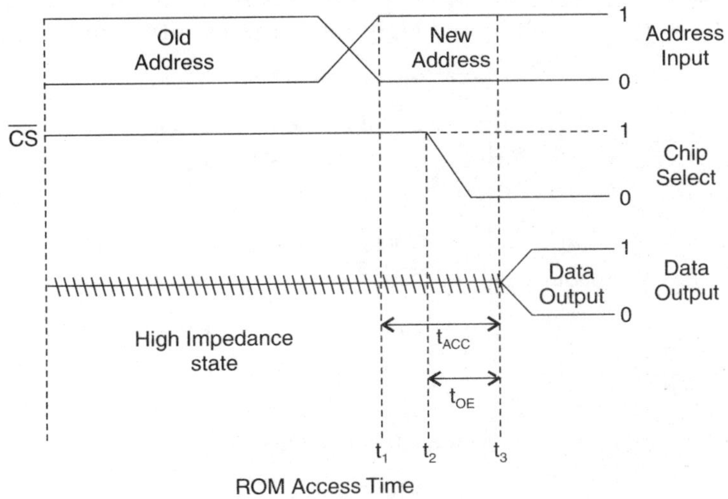

Fig. 7.13

The waveform of address inputs shows the switch over from old address to new valid address. The waveforms of active low chip select ($\overline{CS}$) and data output are also shown.

At time to the address inputs are at some specific level, some High and some Low. $\overline{CS}$ is High and ROM data outputs are in High impedance state (shown by hatched line). Just before time t_1, the address inputs are changing for a new read operation.

At time t_1, the new address is valid. The process of decoding of new address inputs (to select the register from which data is to be read) begins at time t_1.

At time t_2, $\overline{CS}$ goes from High to Low (*i.e.*, $\overline{CS}$ is activated).

At time t_3, the output changes from high impedance state to valid state (*i.e.*, output becomes valid). The time delay from t_1 to t_3 is the ROM access time (t_{ACC}). Typical values of ROM access time are 30*n*S to 90*n*S for bipolar ROM and 35–500*n*S for *n*MOS ROM.

The time delay from t_2 and t_3 is called output enable time t_{OE} (this is time delay between $\overline{CS}$ input and valid data output). This time is about 10*n*S to 20*n*S for Bipolar ROM at 25–100*n*S for *n*MOS ROMs.

7.6 FLASH MEMORY

Flash memory is a semi-conductor non-volatile memory which, to some extent, combines the advantages of EPROM and EEPROM. Flash memory has high density, low cost, low access time and in circuit electrical erasibility. It has a single transistor EPROM cell. It has thinner gate oxide layer so that high density and electrical erasibility are possible.

The word flash memory comes from rapid erase and write times.

Some flash memories use bulk erase operation so that all cells on the clip can be erased simultaneously in about 100 ms or so.

Some other flash memories have sector erasure *i.e.*, specific sectors of memory can be erased as per requirement.

IC 28F256A is a CMOS flash memory chip having a capacity of 32 K × 8. It has 15 address inputs (2^{15} = 32768, *i.e.*, 32 k) and eight data inputs/output pins for 8 bit data).

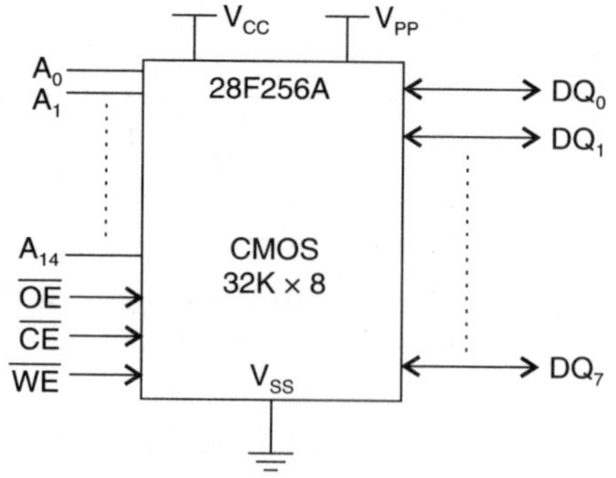

Flash Memory Chip

Fig. 7.14

	Inputs			
Mode	$\overline{CE}$	$\overline{OE}$	$\overline{WE}$	State
Read	Low	Low	High	Data Out
Standby	High	×	×	High Impedance
Write	Low	High	Low	Data In

Status of Control Input

Advantages of ROM:

(*i*) It is cheaper than RAM.

(*ii*) It is non-volatile. Therefore, the contents are not lost when power is switched off.

(*iii*) It is available in larger sizes than RAM.

(*iv*) Its contents are always known and can be easily tested.

(*v*) It does not require refreshing.

(*vi*) There is no chance of any accidental change in its contents.

Advantages of RAM:

(*i*) It can be updated and replaced.

(*ii*) It can serve as temporary data storage.

(*iii*) It does not require lead time (as in ROM) or programming time (as in PROM).

(*iv*) It does not require any programming equipment.

7.7 CACHE MEMORY

Cache memory is a small high speed memory between CPU and main memory. When cache memory is used some information stored in primary and secondary storage area is transferred to cache when the CPU is processing the information in primary storage. When CPU has processed the information in primary storage, cache memory contents are transferred at high speed to primary storage. During the system operation recently used data and instructions are called **held in cache memory.** The speed of cache memory may be about four to five times the speed of the main memory. Its size varies from about 16 K bytes to about 256 K bytes.

7.8 PROGRAMMABLE LOGIC DEVICE

It is an IC which contains a large number of interconnected gates, flip-flops and registers. Many of these connections are through fusible links which can be broken.

The IC is programmable because the specific function of IC for a given application is determined by the selective breaking of some of the inter-connections while leaving the others intact. This process of selective breaking is done by the manufacturer to meet the requirements of the user or can be done by the user also. Here, there is two planes one is AND plane and other is OR plane.

Advantages of PLD:

(*i*) High density (and low space requirement).

(*ii*) Low production cost.

(*iii*) Lower power requirement.

(*iv*) More flexibility to design.

(*v*) Easy programming.

Classification of PLD:

(*i*) Programmable Read Only Memory (PROM).

(*ii*) Programmable Array Logic (PAL).

(*iii*) Programmable Logic Array (PLA).

(*iv*) Field Programmable Gate Array (FPGA).

(*v*) Generic Logic Array (GLA).

7.8.1 PROM as PLD

In the PROM architecture for PLD, the AND array is hardwired but the OR array is programmable. An intact fuse represents/while a blown fuse represents '0'.

PROM can generate any logic function of the input variables because it generates every possible AND product term. Therefore, it is useful in cases where it is required that every input combination should be available. When the number of input variables is large a PROM becomes impractical because the number of fuses double for every one increase in input variables.

Advantages of PROM as PLD:

(*i*) Less cost.

(*ii*) No simplification of logic function is required.

(*iii*) Flexibility in design because the design can be modified easily.

Disadvantages of PROM as PLD:

(*i*) The complete circuit is not utilised.

(*ii*) More power requirement.

(*iii*) Not suitable when the number of input variables is large.

Example:

Design $Y_0 = A, Y_1 = C\bar{B}\,\bar{A}, Y_2 = C\bar{B}A$ using PROM.

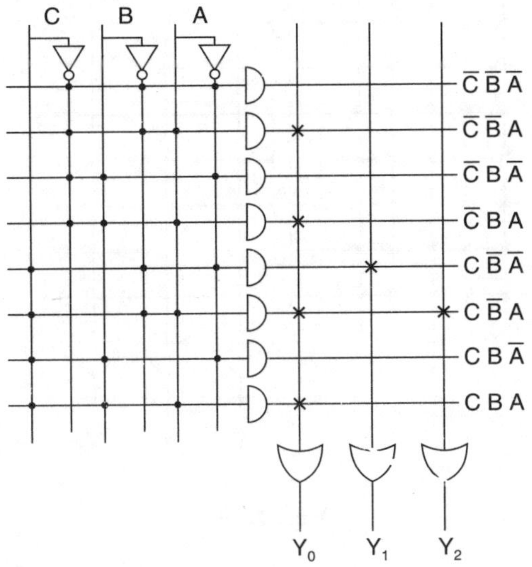

Fig. 7.15

Here No. of output = 3 . (Y_0 , Y_1, Y_2).

Three variables are used. So, the AND Gate is required = $2^3 = 8$.

$$Y_0 = A = A(B + \bar{B})(C + \bar{C})$$

$$= (AB + A\bar{B})(C + \bar{C})$$

$$= ABC + A\bar{B}C + AB\bar{C} + A\bar{B}\,\bar{C} = CBA + C\bar{B}A + \bar{C}BA + \bar{C}\,\bar{B}A.$$

7.8.2 Programmable Array Logic (PAL)

Here AND array is programmable and OR array is fixed.

Example : Design the circuit by PAL.

$$Y_0 = A$$

$$Y_1 = C\bar{B}\,\bar{A}$$

$$Y_2 = C\bar{B}A$$

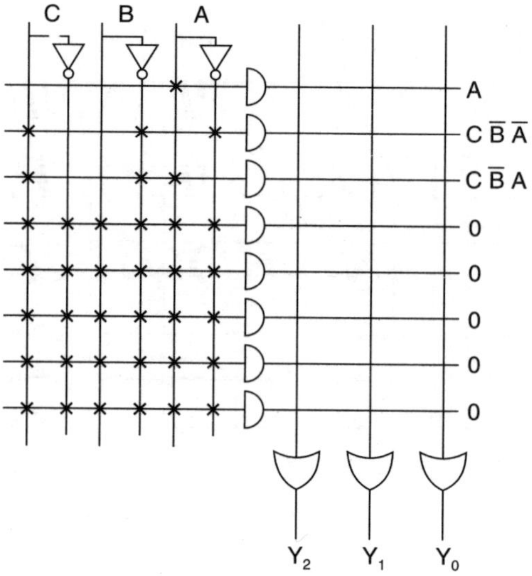

Fig. 7.16

7.8.3 Programmable Logic Array (PLA)

Here AND and OR plane both are programmable.

$$Y_0 = A$$

$$Y_1 = C\,\overline{B}\,\overline{A}$$

$$Y_2 = C\,\overline{B}\,A$$

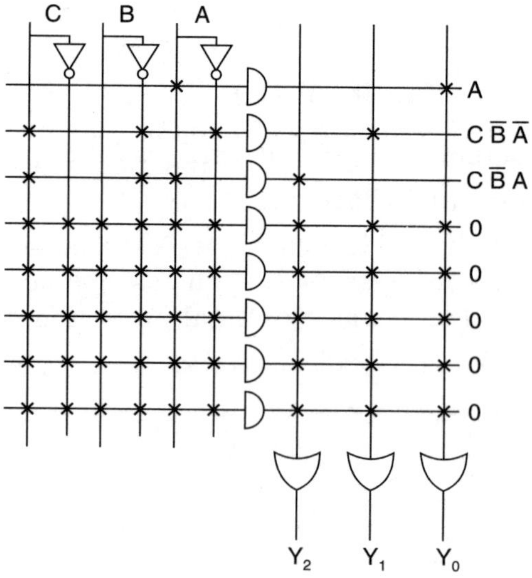

Fig. 7.17

Chapter 8

FLOORPLANNING, PLACEMENT, PARTITIONING AND ROUTING

8.1 FLOORPLANNING

Floorplanning is the art of any physical design.

A well thought-out floorplan leads to an ASIC design with higher performance and optimum area.

Floorplanning can be challenging in that it deals with the placement of I/O pads and macros as well as power and ground structures.

Before one proceeds with physical floorplanning one needs to make sure that the data used during the course of physical design activity is prepared properly.

Proper data preparation is essential to allASIC physical designs in order to implement a correct-by-construction design.

The types of data that are required to start a physical design are:

- Related technology and library files
- Circuit description of the design in the form of netlist representation
- Timing requirements or design constraints
- Floorplan

8.2 TECHNOLOGY FILE

Technology files contain information or commands that are used to configure structures, parameters (such as physical design rules and parasitic extractions), and limits of an ASIC design targeted to specific process technology.

Once the initial technology file is created, several trial runs need to be performed and results, such as standard cell placement, routing quality, and accuracy of parasitic extraction, should be carefully analyzed. Based upon the final inspection, technology files may require further refinement for optimal performance.

Technology rule basics are as follows:

- Manufacturing grid
- Routing grid
- Standard cell placement tile
- Routing layer definition
- Placement and routing blockage layer definition

- Via definition
- Conducting layer density rule
- Metal layer slotting rule
- Routing layer physical profile
- Antenna definition

Manufacturing grid is determined by the smallest geometry that a semiconductor foundry can process.

All drawn geometries during physical design must snap to this manufacturing grid.

Routing grids or tracks are used by physical synthesis and place-and-route tools during detail routing.

The routing tracks can be grid-based, gridlessbased, or subgrid-based.

Standard cell placement tile is used during the placement phase.

The placement tile is defined by one vertical routing track and the standard cell height.

Routing layer definition is used to define the layers that are used to route the design. These definitions include wire width, routing pitch, and preferred routing direction such as vertical, horizontal, or diagonal.

Placement and routing blockage layer definitions are internal to physical design tools and are used to define "keep-out" regions for standard cell placement and routing.

Via definition defines the layer, size, and type for connection between overlapping geometries of conductor for different conductive layers. This cut layer, or via, can be a single via, stacked via, or array of via.

Conducting layer density rule defines the percentage of area of the chip that is required for processes that are using Chemical Mechanical Polishing (CMP) for each physical layer in the design.

Configuration of metal layers slotting rule defines the minimum layer width that may need to have slotting features (*i.e.,* a cut inside a wide routing layer).

This rule varies between foundries and is used to limit mechanical stress for a given conducting layer.

Physical profile for each layer is used to define and include conductor thickness, height, and interlayer dielectric thickness.

Definition of the electrical interconnect profile includes resistance and dielectric constants.

Antenna definition for each layer configures the physical design tools for automatic antenna repair. Antenna phenomena occur during the metallization process when some wires connected to the polysilicon gates of transistors are left unconnected until the upper conducting layers are deposited.

A long wire connected to the gate of MOSFT can act as a capacitor or antenna that collects charges during the plasma-etching step.

8.3 CIRCUIT DESCRIPTION

In the early days of ASIC design, logic designers used schematic capture, or circuit editing tools, to implement the design.

Once the design was captured, the circuit description was imported from capturing tools in some sort of format for physical design.

During this time, the rapid emergence of Computer Aided Engineering (CAE) along with Computer Aided Design (CAD) technology led to the development of a broad range of interchangeable data formats and hardware description languages.

Unfortunately, most of these circuit descriptions were limited to specific companies and data types.

ASIC and physical designers who wished to use a combination of different CAD tools were forced to convert among various formats in order to complete the design. This cumbersome and timeconsuming translation process drove the need for a standard electronic design interchange format.

To address this issue, the first electronic industry standard format (Electronic Design Interchange Format, or EDIF) was introduced.

EDIF is very rich in format and is capable of representing connectivity information, schematic drawings, technology and design rules, and Multi Chip Module (MCM) description, as well as allowing transfer of documentation association with physical layouts.

During the same time, two major formats were introduced-Very High Speed Integrated Circuit (VHSIC) or Hardware Description Language (VHDL) and the Verilog description language (commonly referred to as Verilog).

From an ASIC design perspective, VHDL is not a circuit design tool but merely creates an accurate model of circuit designs. This language, however, has been a phenomenal success because it has the utility and ability to meet circuit and system design requirements.

VHDL is independent of design tools and is a powerful language for the modeling of hardware timing.

After building the netlist we will get the appropriate gate level architecture.

In the corresponding schematic presentation there are two module definitions — **bottom_level** and **top_level** and their corresponding input and output ports. The structural module **top_level** has one instance of **bottom_level.** In addition, both **top_level** and **bottom_level** modules contain gate types (*e.g.,* OR_type, AND_type, TRIBUF_type, and BUF_type) that are considered cell types. These cell types are built into a standard cell library and have predefined functions.

After processing the imported netlist, the next step is to apply design Constraints.

8.4 DESIGN CONSTRAINTS

Design constraints are ASIC design specifications that are applied during logic and physical synthesis.

Each tool attempts to meet two general design constraints :

- Timing constraints
- Design rule constraints

Timing constraints are user-specified and are related to speed, area, and the power consumption of the ASIC design.

Timing constraints utilized by physical design tools are performance related.

The most basic timing constraints are as follows:

- System clock definition and clock delays
- Multiple cycle paths
- Input and output delays
- Minimum and maximum path delays
- Input transition and output load capacitance
- False paths.

System clocks, and their delays, are extremely important constraints in ASIC designs. System clocks are typically supplied externally, but can be generated inside an ASIC. All delays, especially in a synchronous ASIC design, are dependent upon the system clocks.

Multiple cycle paths are for ASIC designs that have a non-single cycle clock timing requirement. This directs the physical design tools to avoid optimization of data paths that have non-single clock behaviour.

Input and output delays are used to constrain the boundary of external paths in an ASIC design. These constraints specify point-to-point delays from external inputs to the first registers and from registers to the outputs of an ASIC design.

Minimum and maximum path delays provide greater flexibility for physical synthesis tools that have a point-to-point optimization capability. This means that one can specify timing constraints from one specific point (*e.g.,* pin or port) in the ASIC design to another, provided such a path exists between the two specified points.

Input transition and output capacitance loads are used to constrain the input slew rate and output capacitance of an ASIC device input and output pins.

These constraints have a direct effect on the final ASIC design timing.

False paths are used to specify point-to-point non-critical timing either internal or external to an ASIC design. Properly identifying these non-critical timing paths has a great impact on physical design tools' performance.

Design rule constraints have precedence over timing constraints because they have to be met in order to realize a functional ASIC design.

There are four types of major design rule constraints:

- Maximum number of fan-outs
- Maximum transitions
- Maximum capacitance
- Maximum wire length

Maximum number of fan-outs specify the number of destinations that one cell can connect to for each standard cell in the library. This constraint can also be applied at the ASIC design level during the physical synthesis to control the number of connections one cell can make.

Maximum transition constraint is the maximum allowable input transitions for each individual cell in the standard cell library. Apart from each element in the standard cell library, this constraint can be applied to a specific net or to an entire ASIC design.

Maximum capacitance constraint behaves similarly to maximum transition constraint, but the cost is based on the total capacitance that a particular standard cell can drive any interconnection in the ASIC design. It should be noted that this constraint is fully independent of maximum transition, and therefore, it can be used in conjunction with maximum transition.

Maximum wire length constraint is useful for controlling the length of wire to reduce the possibility of two parallel long wires of the same type. Parallel long wires of the same type may have a negative impact on the noise injection and may cause crosstalk.

These design rule constraints are mainly achieved by properly inserting buffers at various stages of physical design. Thus, it is imperative to control the buffering in an ASIC design during place-and-route to minimize area impact.

8.5 DESIGN PLANNING

Efficient design implementation of any ASIC requires an appropriate style or planning approach that enhances the implementation cycle time and allows the design goals such as area and performance to be met.

There are two style alternatives for design implementation of an ASIC-**flat or hierarchical.**

For small to medium ASIC's, flattening the design is most suited; for very large and/or concurrent ASIC designs, partitioning the design into subdesigns, or hierarchial style, is preferred.

The flat implementation style:

The flat implementation style provides better area usage and requires effort during physical design and timing closure compared to the hierarchical style.

The area advantage is mainly due to there being no need to reserve extra space around each subdesign partition for power, ground and resources for the routing.

Timing analysis efficiencies arise from the fact that the entire design can be analyzed at once rather than analyzing each subcircuit separately and then analyzing the assembled design later.

The disadvantage of this method is that it requires a large memory space for data and run time increases rapidly with design size.

The hierarchical implementation style:

The hierarchical implementation style is mostly used for very large and/or concurrent ASIC designs where there is a need for a substantial amount of computing capability for data processing.

In addition, it is used when subcircuits are designed individually.

However, hierarchical design implementation may degrade the performance of the final ASIC.

This performance degradation is mainly because the components forming the critical path may reside in different partitions within the design thereby extending the length of the critical path.

Therefore, when using a hierarchical design implementation style one needs to assign the critical components to the same partition or generate proper timing constraints in order to keep the critical timing components close to each other and thus minimize the length of the critical path within the ASIC.

In the hierarchical design implementation style, an ASIC design can be partitioned **logically** or **physically.**

Logical partitioning:

Logical partitioning takes place in the early stages of ASIC design (*i.e.,* RTL coding).

The design is partitioned according to its logical functions, as well as physical constraints, such as interconnectivity to other partitions or subcircuits within the design.

In logical partitioning, each partition is place-and-routed separately and is placed as a macro, or block, at the ASIC top level.

Physical partitioning:

Physical partitioning is performed during the physical design activity.

Once the entire ASIC design is imported into physical design tools, partitions can be created which combine several subcircuits, or a large circuit can be partitioned into several subcircuits.

Most often, these partitions are formed by recursively partitioning a rectangular area containing the design using vertical or horizontal cut lines.

Physical partitioning is used for minimizing delay (subject to the constraints applied to the cluster or managing circuit complexity) and satisfying timing and other design requirements in a small number of subcircuits.

Initially, these partitions have undefined dimensions and fixed area (*i.e.,* the total area of cells or instance added to the partition) with their associated ports, or terminals, assigned to their boundaries such that the connectivity among them is minimized. In order to place these partitions, or blocks, at the chip level, their dimensions as well as their port placement must be defined.

Method:

One method that is suggested to estimate the perimeter of a macro instance is to use the number of terminals or ports allowed for each block and their associated spacing between each terminal.

The relationship between the perimeter of each partition and the number of associated terminals is given by

$$P = NS \qquad ...(8.1)$$

where P is the perimeter of the physical partition block, N is the number of terminals or ports, and S corresponds to spacing between terminals.

The aspect ratio bounds that are generated by the hierarchical floorplan algorithm must have the flexibility to ensure that each macro instance shape can be reshaped for optimum placement.

Because one requirement for an ASIC design is to fit into the smallest available die size, during the partitioning process for each partition several layout alternatives are considered by modification of the dimensions and terminal placements along the boundary of each partition in the design such that the amount of unused and routing area between each partition is minimized.

Regardless of the physical design implementation style, after physical database creation using the imported netlist and corresponding library and technology files, the first step is to determine ASIC device and core width and height. In addition, standard cell rows and I/O pad sites are created.

The rows and I/O pad sites are for standard cell and I/O pad placement. Fig. 8.1 shows an initial ASIC design floorplan.

The height of a row is equal to the height of the standard cells in the library.

If there are any multiple-height standard cells in the library, they will occupy multiple rows.

Most of the time, standard rows are created by abutment.

The standard rows are oriented in alternating 180-degree rotation or are flipped along the X-axis so that the standard cells can share power and ground busses.

If the ASIC core has routing congestion owing to the limited number of routing layers, one solution is to create routing channels between rows.

These all can be separated individually or as pairs.

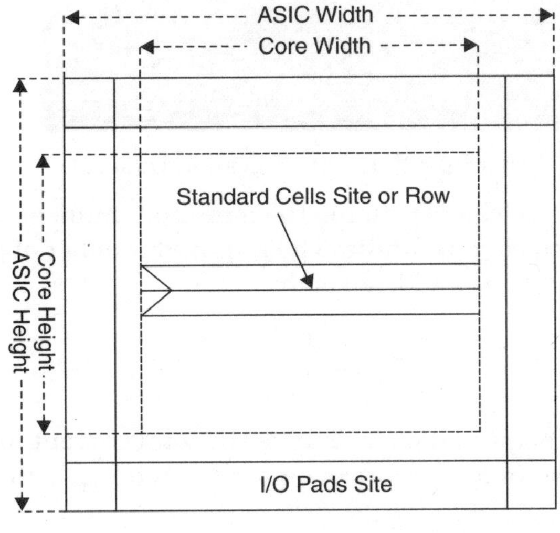

Fig. 8.1. ASIC Design Initial Floorplan

8.6 PAD PLACEMENT

Correct I/O pad placement and selection is important for the correct function of any ASIC design.

For a given ASIC design there are three types of I/O pads. These pads are power, ground, and signal.

It is critical to functional operation of an ASIC design to insure that the pads have adequate power and ground connections and are placed properly in order to eliminate electromigration and current-switching noise related problems.

Electromigration (EM) is the movement or molecular transfer of metal from one area to another area that is caused by an excessive electrical current in the direction of electron flow (electron "wind").

Electromigration currents exceeding recommended guidelines can result in premature ASIC device failure.

Exceeding electromigration current density limits can create voids or hillocks, resulting in increased metal resistance, or shorts between wires, and can impair ASIC performance.

Figure 8.2 shows electromigration damage due to excess current captured by Electron Scanning Microscopy (ESM) with 10 K magnification.

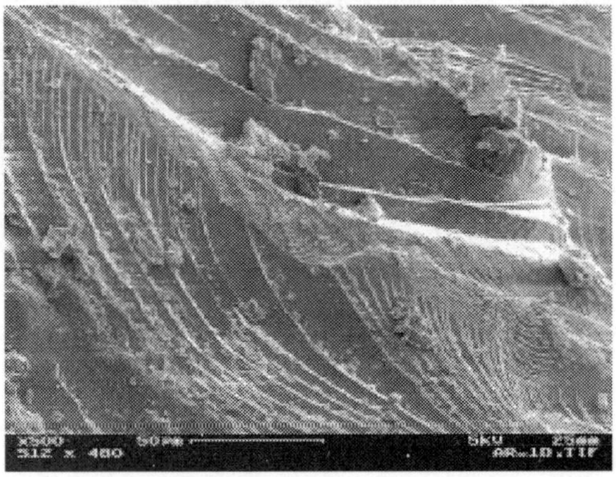

Fig. 8.2. Electromigration Damage

Using equation 8.2 one can determine the minimum number of ground pads required to satisfy the electromigration current limits. The required number of power pads is equal to the required number of ground pads and is given by

$$N_{gnd} = \frac{I_{total}}{I_{max}} \qquad ...(8.2)$$

where N_{gnd} is the number of ground pads ; I_{total} is the total current in an ASIC design (which is the sum of its static and dynamic currents in amperes); and I_{max} is the maximum EM current in amperes per ground pad.

Switching noise is generated when ASIC outputs make transitions between states.

There are two types of mechanisms that can cause noise:

- $\frac{dv}{dt}$ caused by a capacitive coupling effect
- $\frac{di}{dt}$ caused by an inductive switching effect

The capacitive coupling effect is the disturbance on the adjacent package pin caused when switching transients inject pulses via parasitic coupling capacitance.

To reduce or eliminate capacitive coupling effects during I/O pad placement and selection, one may consider the following guidelines:

- Isolate sensitive asynchronous inputs such as clock or bidirectional pins from other switching signal pads with power or ground pads.
- Group bidirectional pads together such that all are in the input or output mode.
- Group slow input pads together positioning them on higher inductance package pins.
- Use input pads with hysteresis as much as possible.

The inductive switching effect is related to simultaneous switching ASIC outputs that induce rapid current changes in the power and ground busses.

The inductance in the power and ground pins cause voltage fluctuations in the ASIC internal power and ground level relative to the external system.

Successful reduction of inductive switching noise can be accomplished by the following:

- Reduce the number of outputs that switch simultaneously by dividing them into groups with each group having a number of delay buffers inserted into their data paths.
- Use the lowest rated sink current or low-noise output pads as long as speed is not an issue.
- Place the simultaneously switching output or bidirectional pads together and distribute power and ground pads among them according to their relative noise rating.
- Assign static and low frequency input pads to higher inductance package pins.
- Reduce the effective power and ground pin inductance by assigning as many power and ground pads as possible.

8.7 POWER PLANNING

The next step is to plan and create power and ground structures for both I/O pads and core logic.

The I/O pads' power and ground busses, are built into the pad itself and will be connected by abutment.

For core logic, there is a core ring enclosing the core with one or more sets of power and ground rings. A horizontal metal layer is used to define the top and bottom sides, or any other horizontal segment, while the vertical metal layer is utilized for left, right, and any other vertical segment.

These vertical and horizontal segments are connected through an appropriate via cut.

The next consideration is to construct the standard cell power and ground that is internal to the core logic.

These internal core power and ground busses consist of one or two sets of wires or strips that repeat at regular intervals across the core logic, or specified region, within the design.

Each of these power and ground strips run vertically, horizontally, or in both directions.

Figure 8.3 illustrates these types of power and ground connections.

If these strips run both vertically and horizontally at regular intervals, then the style is known as power mesh. The total number of strips and interval distance is solely dependent on the ASIC core power consumption.

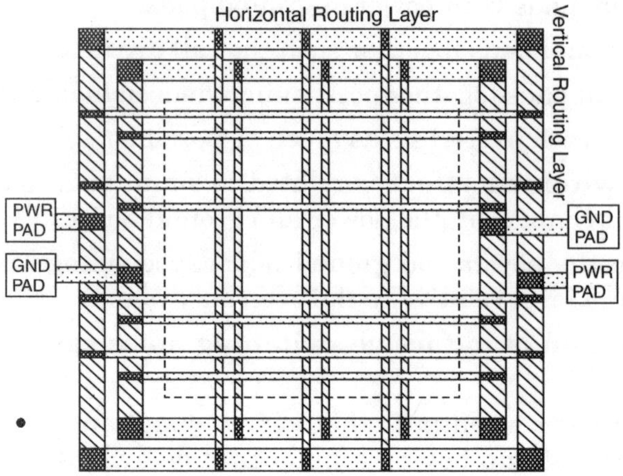

Fig. 8.3. Ring and Core Power and Ground

As the ASIC core power consumption (dynamic and static) increases, the distance of power and ground strip intervals increases. This increase in the power and ground strip intervals is used mainly to reduce overall ASIC voltage drop, thereby improving ASIC design performance.

In addition to the core power and ground ring, macro power and ground rings need to be created using proper vertical and horizontal metal layers. A macro ring encloses one or more macros, completely or partially, with one or more sets of power and ground rings.

Another important consideration is that when both analog and digital blocks are present in an ASIC design, there is a need for special care to insure that there is no noise injection from digital blocks or core into the sensitive circuits of analog blocks through power and ground supply connections.

The most effective method is to decouple the digital and analog power and ground by routing the digital power/ground (DP and DG) and analog power/ground (AP and AG) supply connections separately as shown in Fig. 8.4.

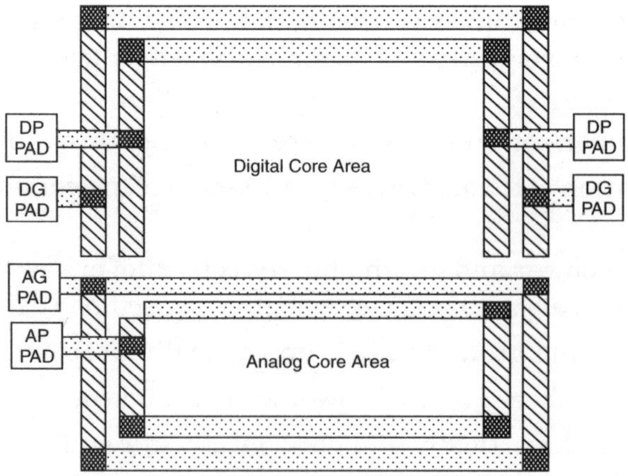

Fig. 8.4. Decoupled Analog and Digital Core Power Supply

However, this decoupling will not be completed, if there is ground connectivity from the die substrate through standard cells' ground (*i.e.,* source).

Depending on the ASIC design's power supply requirements, the width of the core macro power and ground ring could exceed the maximum allowable in order to reduce their resistance during floorplanning and power analysis.

This increase in power and ground width could be problematic from manufacturing point of view.

The main problem with wide metal (*i.e.,* exceeding manufacturing limits) is that a metal layer cannot be processed with uniform thickness, especially when applied over a wide area.

The metal becomes thin in the middle and thick on the edges causing yield and current density problems.

To solve this metal density problem, one can either use multiple power and ground busses or use metal slotting.

8.8 MACRO PLACEMENT

Typically, the placement of macros takes place after I/O placement, and after power and ground bus structure has been defined.

Macros may be memories, analog blocks, or in the case of hierarchical style, an individually placed and routed subcircuit.

Proper placement of these macros has a greater impact on the quality of the final ASIC design.

Macro placement can be manual or automatic.

Manual macro placement is more efficient when there are few macros to be placed and their relationship with the rest of the ASIC design is known.

Automatic macro placement is more appropriate if there is not enough information on which to base the initial macro placement and/or the number of macros is large.

During the macro placement step, one needs to make sure that there is enough area between blocks for interconnections.

This process (commonly known as channel allocation or channel definition) can be manual or can be accomplished by floorplan tools. The slicing tree is used by the floorplan alogrithm for slicing floorplan during macro placement and to define routing channels between the blocks.

One of the basic algorithms used for automatic macro placement considers that macros are connected to each other by nets and are supposed to exert attractive forces on each other by means of wire length proportional to the distance between these macros.

Automatic macro placement is an iterative process. During the macro placement process, macros are free to move until the equilibrium or optimum wire length is achieved. It is interesting to note that in this algorithm, if there is no relationship between the macros, they tend to repel each other and their placement result may not be optimum.

To avoid area segmentation, macros should be positioned such that the standard cell area is continuous. An area with close to 1 : 1 aspect ratio is recommended as it allows standard cell placers to utilize the area more efficiently and thereby reduce total wire length.

The segmented floorplan leads to an excess of wire length interconnections from the standard cells located at the bottom of the die to those at the top of the die. Thus, it is necessary that the macros be kept along the ASIC core area in order to avoid a floorplan segmentation problem.

Figure 8.5 shows a problematic segmented floorplan that may lead to long interconnections between the bottom and top of the die.

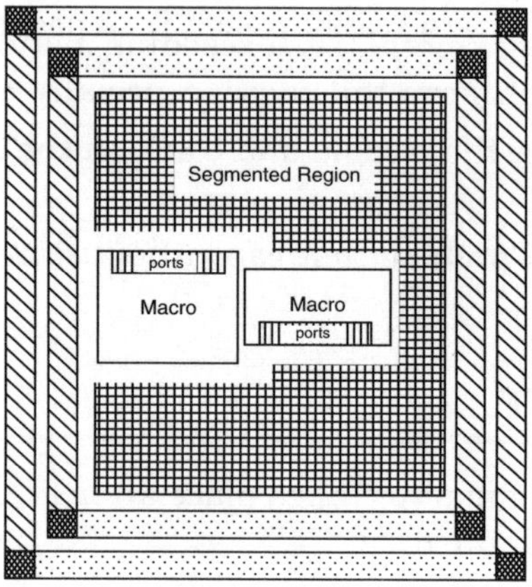

Fig. 8.5. Segmented Floorplan

Another aspect of physical measure is that of macro placements relative to each other, as well as to the standard cell placement, and the macro port accessibility has a direct impact on the chip's final routing.

Figure 8.6 shows a floorplan with macro ports facing the standard cell region, thereby minimizing localized increase in wire length.

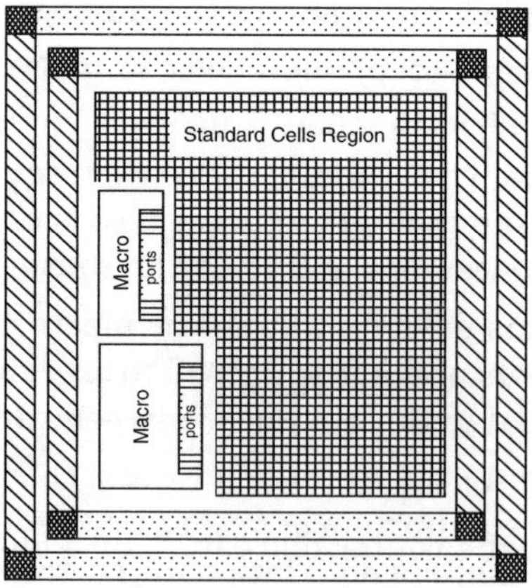

Fig. 8.6. Floorplan with Macros Facing Standard Cells Region

The most common scenarios that cause physical design routing congestion are: there may not be enough space between the macros to provide routing channels — especially for I/O connections and macros (if these macros are placed at the ASIC periphery and over the macro); routing is prohibited; or standard cell trap pockets may be around the edges of macros or within the corners of the floorplan.

Standard cell trap pockets are long, thin channels between macros. If many cells are placed in these channels, routing congestion can result. Therefore, these channels need to be kept free for most standard cells and should be available for repeater or buffer insertion (if this type of insertion is supported by the physical design tool).

Figure 8.7 shows a floorplan with a standard cell trap pocket.

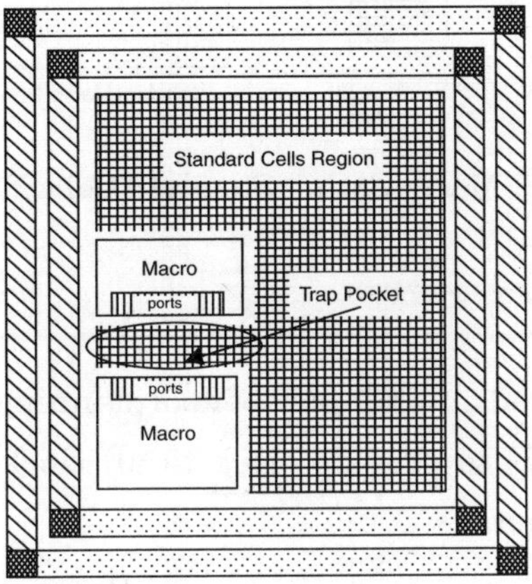

Fig. 8.7. Floorplan with Standard Cells Trap Pocket

After macro placement and before performing global routing, most physical designs require keep-out or buffer-only regions to be defined by drawing a blockage layer over an area containing macros to prevent the placer from moving any standard cells into those regions.

Naturally, the wires that are used in keep-out regions have a tendency to be long. By allowing buffer insertion in those areas by using a buffer-only region (or blockage), the placer will taper these long nets and thus avoid the long transition times associated with them.

These blockage layers are created over pre-placed macros such that their power and ground rings are covered.

Blockage layers are also used to relieve routing congestion around the macro's corners. When a macro is blocked on many routing layers, wires have a tendency to detour around corners and connect to nearby standard cells thereby creating routing congestion at the corner of the macros.

To reverse more resources for the router, one can draw a blockage layer at these corners. These blockage regions can be simple or gradual as illustrated in Fig. 8.8.

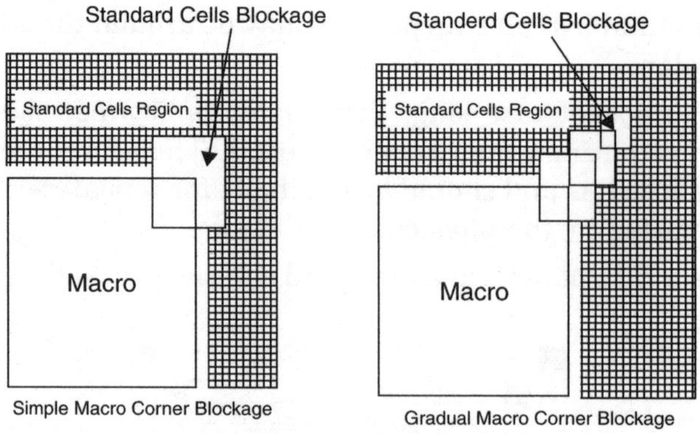

Fig. 8.8. Macro Corner Standard Cells Blockage

8.9 CLOCK PLANNING

Clock planning is considered another part of floorplanning. The idea of the implementation of clock distribution networks is to provide clock to all clocked elements in the design in a symmetrically-structured manner.

Although most ASIC designs use clock tree synthesis, clock tree synthesis may not be sufficient for very high-performance and synchronized designs.

In this case, one needs to implement the distributed clock networks manually in order to minimize the skew between communicating elements due to their line resistance and capacitance.

The basic idea of manual implementation of clock distribution networks is to build a low resistance/capacitance grid similar to power and ground mesh that covers the entire logic core area as shown in Fig. 8.9.

It is important to note that this type of clock grid does not rely on the matching of components such as clock buffers. However, it can present a systematic clock skew.

In order to minimize such clock skew, a clock tree that balances the rise and fall time at each clock buffer node should be utilized during clock planning.

This minimizes the hot-electron effect. The problem of hot-electron occurs when an electron gains enough energy to escape from a channel into a gate oxide.

The presence of hot-electron effect in the gate oxide area causes the threshold voltage of the device to change and thus alters the delay of the clock buffers.

This in turns leads to an additional skew. Hence, balancing the rise and fall times of all clock buffers means that hot-electron effect influences the clock buffers at the same rate, minimizing unpredictable skew.

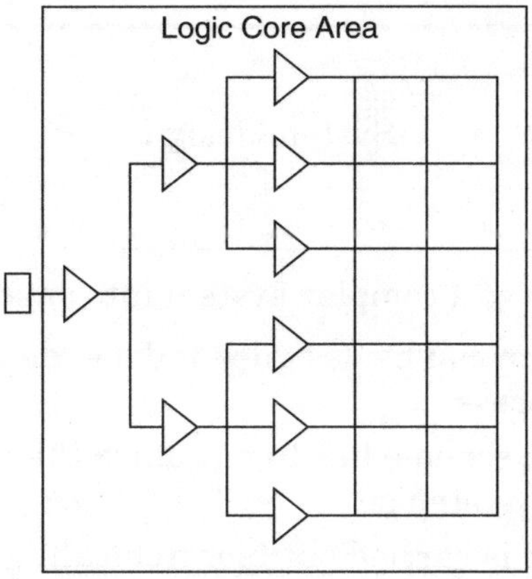

Fig. 8.9. Clock Distribution Network

To minimize the clock skew among all leaf nodes, the clock delay for each sub-block must be determined and the design of the clock planned accordingly.

Figure 8.10 illustrates typical hierarchical clock planning.

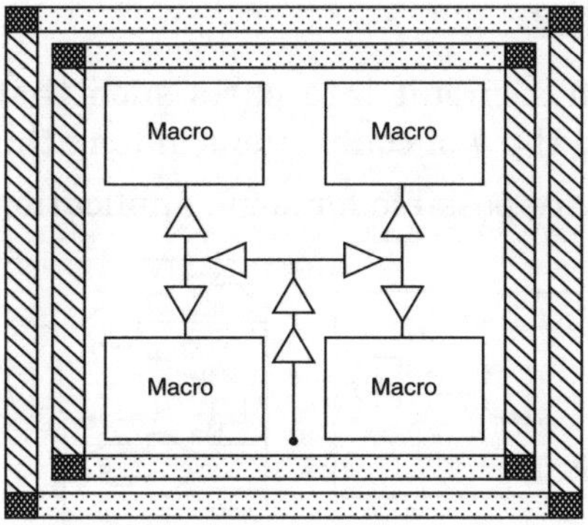

Fig. 8.10. Hierarchical Clock Planning

8.10 PARTITIONING

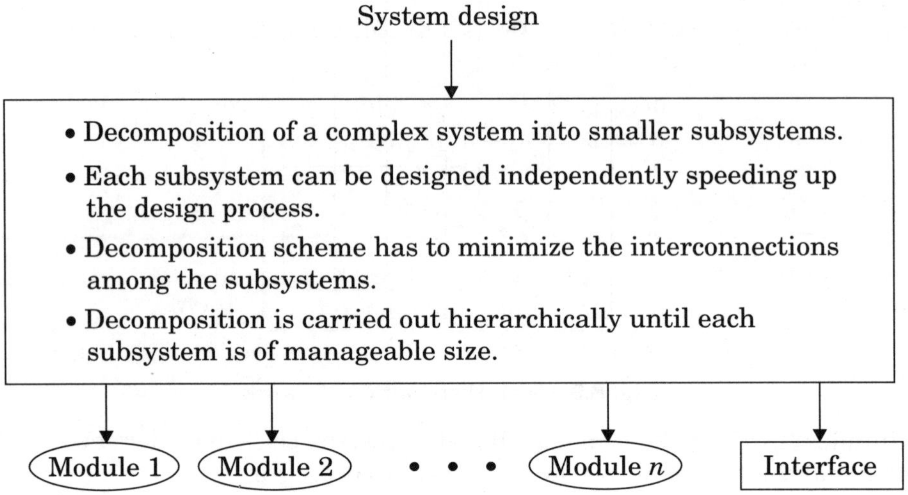

8.10.1 Circuit Partitioning

- **Objective.** Partition a circuit into parts such that every component is within a prescribed range and the # of connections among the components is minimized.

  — More constraints are possible for some applications.

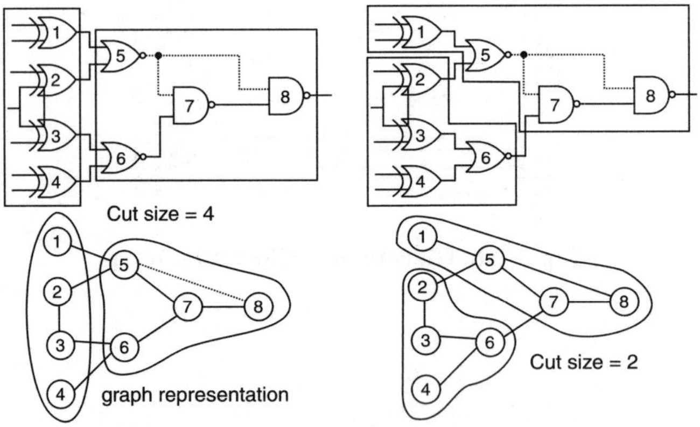

Fig. 8.11

8.11 PROBLEM DEFINITION:PARTITIONING

- ***k*-way partitioning.** Given a graph $G(V, E)$, where each vertex $v \in V$ has a **size** $s(v)$ and each edge $e \in$ E has a **weight** $w(e)$, the problem is to divide the set V into k disjoint subsets $V_1, V_2, \ldots, V_k$, such that an objective function is optimized, subject to certain constraints.

- **Bounded size constraint.** The size of the i-th subset is bounded by B_i ($\Sigma_{v \in V_i} s(v) \leq B_i$).
  — Is the partition balanced ?
- **Min-cut cost between two subsets.** Minimize $\Sigma_{\forall c} = (u, v) \wedge p(u) \neq p(v)\, w(e)$, where $p(u)$ is the partition # of node u.
- The 2-way, balanced partitioning problem is NP-complete, even in its simple form with identical vertex sizes and unit edge weights.

8.12 KERNIGHAN-LIN ALGORITHM

- **Kernighan and Lin,** "An efficient heuristic procedure for partitioning graphs." The Bell System Technical Journal, vol. 49, no. 2, Feb. 1970.
- An **iterative, 2-way, balanced** partitioning (bi-sectioning) heuristic.
- Till the cut size keeps decreasing.
  — Vertex pairs which give the largest decrease or the smallest increase in cut size are exchanged.
  — These vertices are then **locked** (and thus are prohibited from participating in any further exchanges).
  — This process continues until all the vertices are locked.

Kernighan-Lin Algorithm : A Simple Example

- Each edge has a unit weight.

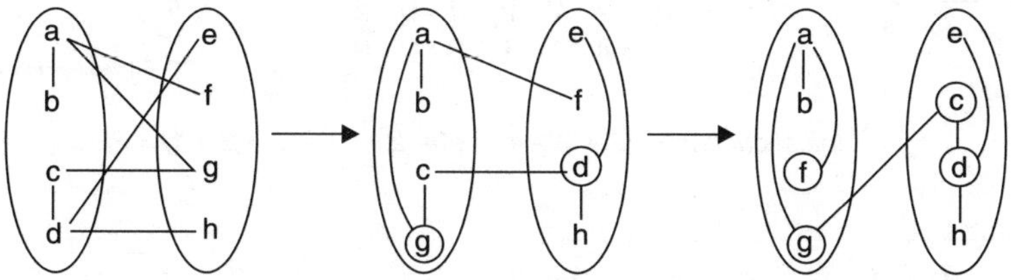

Step #	Vertex pair	Cost reduction	Cut cost
0	–	0	5
1	$\{d, g\}$	3	2
2	$\{c, f\}$	1	1
3	$\{b, h\}$	– 2	3
4	$\{a, e\}$	– 2	5

- Questions : How to compute cost reduction ? What pairs to be swapped ?
  — Consider the change of internal and external connections.

Properties

- Two sets A and B such that $|A| = n = |B|$ and $A \cap B = \phi$.
- **External cost** of $a \in A$; $E_a = \Sigma_{v \in B}\, c_{av}$.

- **Internal cost** of $a \in \mathrm{A} : \mathrm{I}_a = \Sigma_{v \in \mathrm{A}}\ c_{av}$.
- D-value of a vertex a : $\mathrm{D}_a = \mathrm{E}_a - \mathrm{I}_a$ (cost reduction for moving a).
- Cost reduction (gain) for swapping a and b : $g_{ab} = \mathrm{D}_a + \mathrm{D}_b - 2\mathrm{C}_{ab}$.
- If $a \in A$ and $b \in B$ are interchanged, then the new D-values, D′, are given by

$$\mathrm{D}_x' = \mathrm{D}_x + 2c_{xa} - 2c_{xb},\ \forall\, x \in A - \{a\}$$

$$\mathrm{D}_y' = \mathrm{D}_y + 2c_{yb} - 2c_{ya},\ \forall\, y \in B - \{b\}.$$

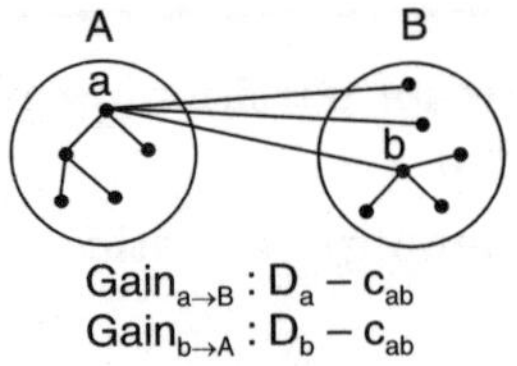

Internal cost vs. External cost

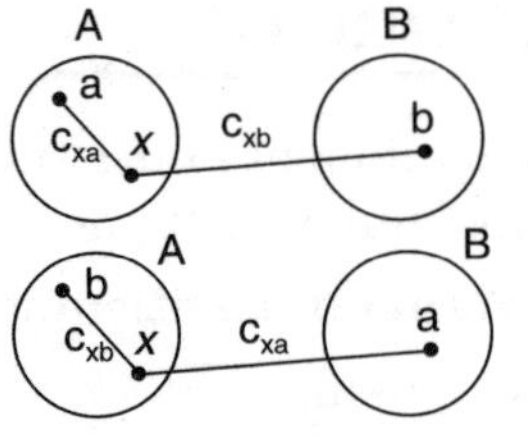

Updating D–values

before swap	*after swap*	C
$-c_{xa}$	$+c_{xa}$	$+2c_{xa}$
$+c_{xb}$	$-c_{xb}$	$-2c_{xb}$

Kernighan-Lin Algorithm : A Weighted Example

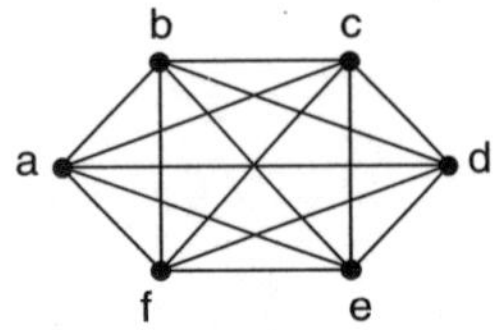

	a	b	c	d	e	f
a	0	1	2	3	2	4
b	1	0	1	4	2	1
c	2	1	0	3	2	1
d	3	4	3	0	4	3
e	2	2	2	4	0	2
f	4	1	1	3	2	0

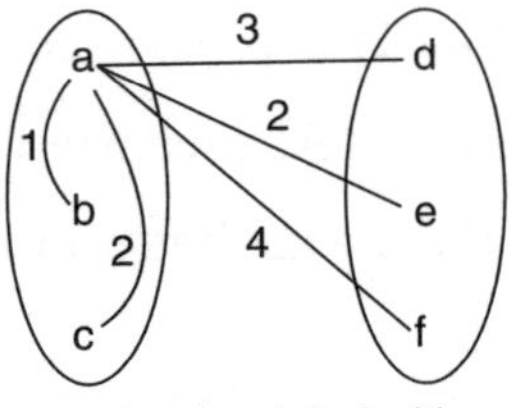

costs associated with a

Initial cut cost = (3 + 2 + 4) + (4 + 2 + 1) + (3 + 2 + 1) = 22

- Iteration I :

$I_a = 1 + 2 = 3$; $E_a = 3 + 2 + 4 = 9$; $D_a = E_a - I_a = 9 - 3 = 6$

$I_b = 1 + 1 = 2$; $E_b = 4 + 2 + 1 = 7$; $D_b = E_b - I_b = 7 - 2 = 5$

$I_c = 2 + 1 = 3$; $E_c = 3 + 2 + 1 = 6$; $D_c = E_c - I_c = 6 - 3 = 3$

$I_d = 4 + 3 = 7$; $E_d = 3 + 4 + 3 = 10$; $D_d = E_d - I_d = 10 - 7 = 3$

$I_e = 4 + 2 = 6$; $E_e = 2 + 2 + 2 = 6$; $D_e = E_e - I_e = 6 - 6 = 0$

$I_f = 3 + 2 = 5$; $E_f = 4 + 1 + 1 = 6$; $D_f = E_f - I_f = 6 - 5 = 1$

Weighted Example (cont'd)

- Iteration 1 :

$I_a = 1 + 2 = 3$; $E_a = 3 + 2 + 4 = 9$; $D_a = E_a - I_a = 9 - 3 = 6$

$I_b = 1 + 1 = 2$; $E_b = 4 + 2 + 1 = 7$; $D_b = E_b - I_b = 7 - 2 = 5$

$I_c = 2 + 1 = 3$; $E_c = 3 + 2 + 1 = 6$; $D_c = E_c - I_c = 6 - 3 = 3$

$I_d = 4 + 3 = 7$; $E_d = 3 + 4 + 3 = 10$; $D_d = E_d - I_d = 10 - 7 = 3$

$I_e = 4 + 2 = 6$; $E_e = 2 + 2 + 2 = 6$; $D_e = E_e - I_e = 6 - 6 = 0$

$I_f = 3 + 2 = 5$; $E_f = 4 + 1 + 1 = 6$; $D_f = E_f - I_f = 6 - 5 = 1$

- $g_{xy} = D_x + D_y - 2c_{xy}$.

$$g_{ad} = D_a + D_d - 2c_{ad} = 6 + 3 - 2 \times 3 = 3$$
$$g_{ae} = 6 + 0 - 2 \times 2 = 2$$
$$g_{af} = 6 + 1 - 2 \times 4 = -1$$
$$g_{bd} = 5 + 3 - 2 \times 4 = 0$$
$$g_{be} = 5 + 0 - 2 \times 2 = 1$$
$$g_{bf} = 5 + 1 - 2 \times 1 = 4\ (maximum)$$
$$g_{cd} = 3 + 3 - 2 \times 3 = 0$$
$$g_{ce} = 3 + 0 - 2 \times 2 = -1$$
$$g_{cf} = 3 + 1 - 2 \times 1 = 2$$

- Swap b and f ! ($\widehat{g_1} = 4$)

Weighted Example (cont'd)

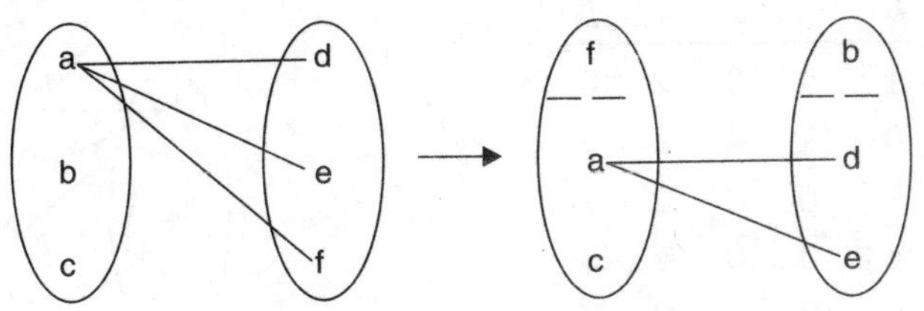

- $D'_x = D_x + 2c_{xp} - 2c_{xq},\ \forall x \in A - \{p\}$ (swap p and q, $p \in A$, $q \in B$)

$$D'_a = D_a + 2c_{ab} - 2c_{af} = 6 + 2 \times 1 - 2 \times 4 = 0$$
$$D'_c = D_c + 2c_{cb} - 2c_{cf} = 3 + 2 \times 1 - 2 \times 1 = 3$$
$$D'_d = D_d + 2c_{de} - 2c_{db} = 3 + 2 \times 3 - 2 \times 4 = 1$$
$$D'_e = D_e + 2c_{ef} - 2c_{ce} = 0 + 2 \times 2 - 2 \times 2 = 0$$

- $g_{xy} = D'_x + D'_y - 2c_{xy}$.

$$g_{ad} = D'_a + D'_d - 2c_{ad} = 0 + 1 - 2 \times 3 = -5$$
$$g_{ae} = D'_a + D'_e - 2c_{ae} = 0 + 0 - 2 \times 2 = -4$$
$$g_{cd} = D'_c + D'_d - 2c_{cd} = 3 + 1 - 2 \times 3 = -2$$
$$g_{ce} = D'_c + D'_e - 2c_{ce} = 3 + 0 - 2 \times 2 = -1\ (maximum)$$

- Swap c and e ! ($\widehat{g_2} = -1$)

Weighted Example (cont'd)

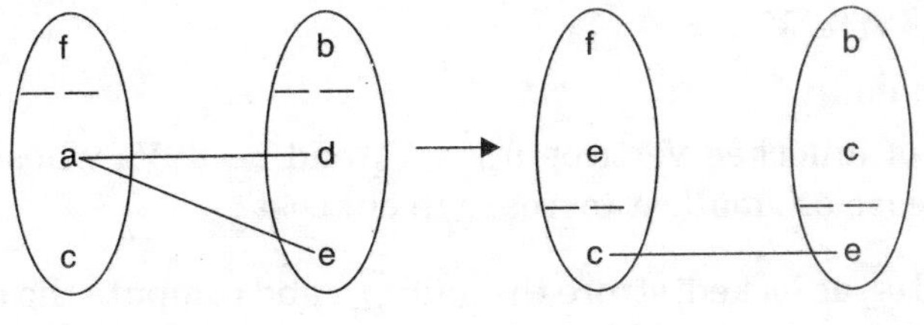

- $D''_x = D'_x + 2c_{xp} - 2c_{xq}, \ \forall x \in A - \{p\}$

$$D''_a = D'_a + 2c_{ac} - 2c_{ae} = 0 + 2 \times 2 - 2 \times 2 = 0$$

$$D''_d = D'_d + 2c_{de} - 2c_{dc} = 1 + 2 \times 4 - 2 \times 3 = 3$$

- $g_{xy} = D''_x + D''_y - 2c_{xy}$

$$g_{ad} = D''_a + D''_d - 2c_{ad} = 0 + 3 - 2 \times 3 = -3 \ (\widehat{g_3} = -3)$$

- Note that this step is redundant ($\sum_{i=1}^{n} \widehat{g_i} = 0$).

- Summary $\widehat{g_1} = g_{bf} = 4$, $\widehat{g_2} = g_{ce} = -1$, $\widehat{g_3} = g_{ad} = -3$.

- Largest partial sum max $\sum_{i=1}^{k} \widehat{g_i} = 4 \ (k = 1)$ $\Rightarrow$ Swap b and f.

Weighted Example (cont'd)

	a	b	c	d	e	f
a	0	1	2	3	2	4
b	1	0	1	4	2	1
c	2	1	0	3	2	1
d	3	4	3	0	4	3
e	2	2	2	4	0	2
f	4	1	1	3	2	0

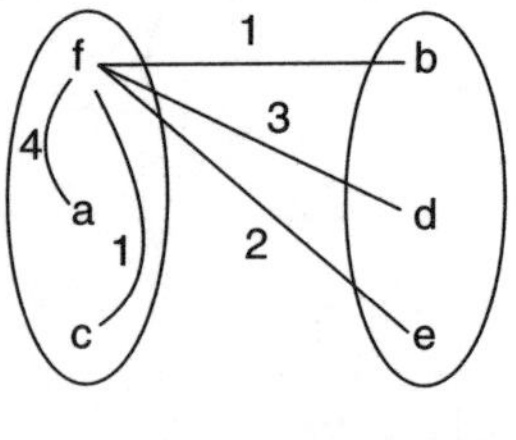

Initial cut cost = (1 + 3 + 2) + (1 + 3 + 2) + (1 + 3 + 2) = 18(22 – 4)

- Iteration 2 : Repeat what we did at Iteration 1 (Initial cost = 22 – 4 = 18).

- Summary : $\widehat{g_1} = g_{ce} = -1$, $\widehat{g_2} = g_{ab} = -3$, $\widehat{g_3} = g_{fd} = 4$.

- Largest partial sum = max $\sum_{i=1}^{k} \widehat{g_i} = 0 \ (k = 3)$ $\Rightarrow$ Stop !

Algorithm : Kernighan-Lin (*G*)

Input : $G = (V, E), |V| = 2n$.

Output : Balanced bi-partition A and B with "small" cut cost.

1. **begin**
2. Bipartition G into A and B such that $|V_A| = |V_B|$, $V_A \cap V_B = \phi$, and $V_A \cup V_B = V$.
3. **repeat**
4. Compute D_v, $\forall v \in V$.
5. **for** $i = 1$ to n do
6. Find a pair of unlocked vertices $v_{ai} \in V_A$ and $v_{bi} \in V_B$ whose exchange makes the largest decrease or smallest increase in cut cost ;
7. Mark v_{ai} and v_{bi} as locked, strore the gain $\widehat{g_i}$, and compute the new D_v, for all unlocked $v \in V$;

8. Find k, such that $G_k = \sum_{i=1}^{k} \widehat{g_i}$ is maximized ;
9. **if** $G_k > 0$ **then**
10. Move v_{a1}, ..., v_{ak} from V_A to V_B and v_{b1}, ..., v_{bk} from V_B to V_A ;
11. Unlock v, $\forall\, v \in V$.
12. **until** $G_k \leq 0$;
13. **end**

Time Complexity

- Line 4 : Initial computation of D : $O(n^2)$
- Line 5 : The **for**-loop : $O(n)$
- The body of the loop : $O(n^2)$
  — Lines 6–7 : Step i takes $(n - i + 1)^2$ time.
- Lines 4–11 : Each pass of the repeat loop : $O(n^3)$.
- Suppose the repeat loop terminates after r passes.
- The total running time : $O(rn^3)$.

Extensions of K-L Algorithm

- **Unequal sized subsets** (assume $n_1 < n_2$)
  1. Partition : $|A| = n_1$ and $|B| = n_2$.
  2. Add $n_2 - n_1$ dummy vertices to set A. Dummy vertices have no connections to the original graph.
  3. Apply the Kernighan-Lin algorithm.
  4. Remove all dummy vertices.
- **Unequal sized "vertices"**
  1. Assume that the smallest "vertex" has unit size.
  2. Replace each vertex of size s with n vertices which are fully connected with edges of infinite weight.
  3. Apply the Kernighan-Lin algorithm.
- ***k*-way partition**
  1. Partition the graph into k equal-sized sets.
  2. Apply the Kernighan-Lin algorithm for each pair of subsets.
  3. Time complexity ? Can be reduced by recursive bi-partition.

A "Better" Implementation of K-L Algorithm

- Sort the D-values in a non-increasing order:

  $D_{\hat{a1}} \geq D_{\hat{a2}} \geq .... \geq D_{\hat{an}}$.

  $D_{\hat{b1}} \geq D_{\hat{b2}} \geq .... \geq D_{\hat{bn}}$.

- Start with a_1, compute $g_{\hat{a1}, \hat{bi}}$, $\forall\, b_i$

  Start with a_2, compute $g_{\hat{a2}, \hat{bi}}$, $\forall\, b_i$

  whenever $D_{\hat{ai}} + D_{\hat{bj}} \leq$ Maximum gain found so far (Quit !).
- Partition $A = \{a, b, c\} : D_a = 6 \,;\, D_b = 5 \,;\, D_c = 3 \,;$

  Partition $B = \{d, e, f\} : D_d = 3 \,;\, D_f = 1 \,;\, D_e = 0 \,;$

  Compute g's

  $g_{ad} = 3 \longrightarrow g_{af} = -1 \longrightarrow g_{ae} = 2$

  $g_{bd} = 0 \longrightarrow g_{bf} = 4 \longrightarrow g_{be} = 1$

  $g_{ad} = 0 \longrightarrow$ No need to compute g_{cf} (Quit !)

  Since $D_c + D_f \leq g_{bf} = 4$.
- Note that the overall time complexity remains $O(rn^3)$.

8.12.1 Drawbacks of the Kernighan-Lin Heuristic

- The K-L heuristic **handles only unit vertex weights.**
  - Vertex weights might represent block sizes, different from blocks to blocks.
  - Reducing a vertex with weight $w(v)$ into a clique with $w(v)$ vertices and edges with a high cost increases the size of the graph substantially.
- The K-L heuristic **handles only exact bisections.**
  - Need dummy vertices to handle the unbalanced problem.
- The K-L heuristic **cannot handle hypergraphs.**
  - Need to handle multiterminal nets directly.
- The **time complexity of a pass is high,** $O(n^3)$.

8.13 COPING WITH HYPERGRAPH

- A hypergraph $H = (N, L)$ consists of a set N of vertices and a set L of hyperedges, where each hyperedge corresponds to a **subset** N_i of distinct vertices with $|N_i| \geq 2$.

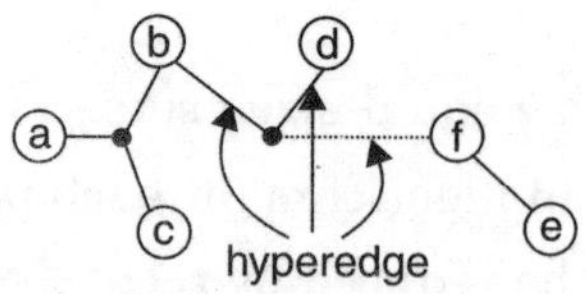

- Schweikert and Kernighan, "A proper model for the partitioning of electrical circuits," 9th Design Automation Workshop, 1972.
- For multiterminal nets, **net cut** is a more accurate measurement for cut cost (*i.e.*, deal with hyperedges).
  - $\{A, B, E\}$, $\{C, D, F\}$ is a good partition.
  - Should not assign the same weight for all edges.

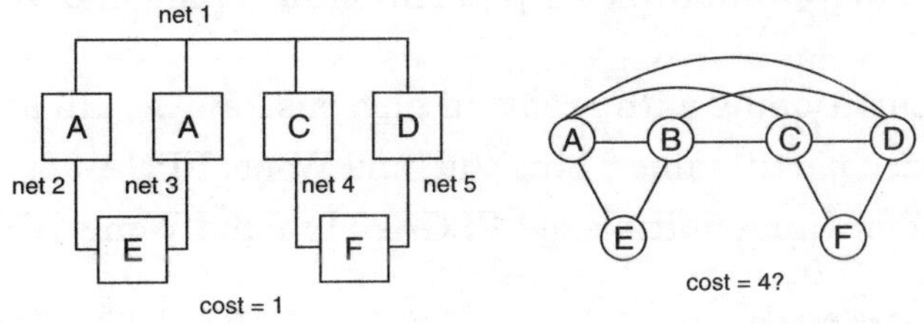

8.14 NET CUT MODEL

- Let $n(i)$ = # of cells associated with Net i.
- Edge weight $w_{xy} = \dfrac{2}{n(i)}$ for an edge connecting cells x and y.

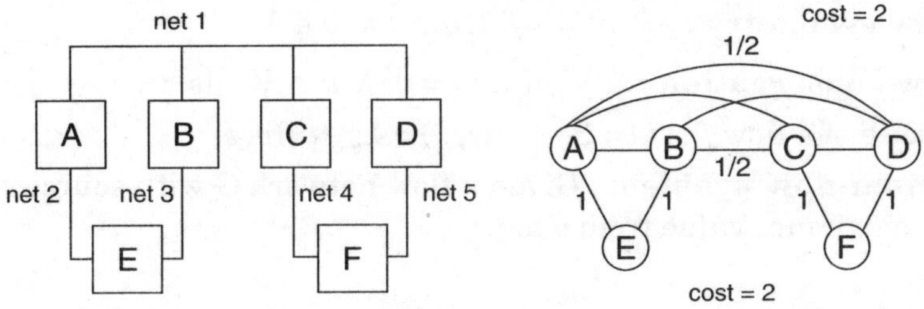

- Easy modification of the K-L heuristic.

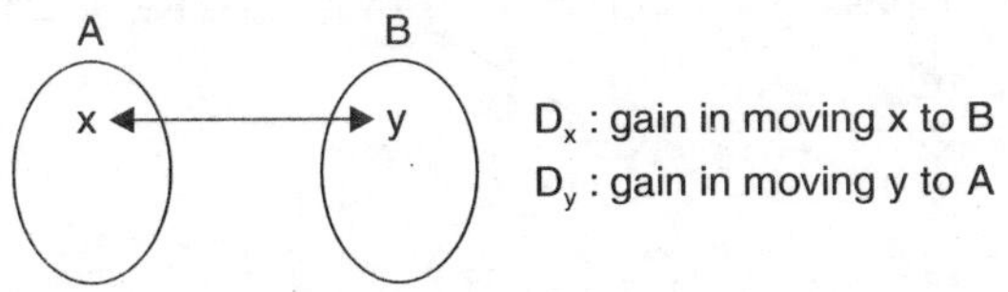

$$g_{xy} = D_x + D_y - \text{Correction}(x, y)$$

8.15 NETWORK FLOW BASED PARTITIONING

- Min-cut balanced partitioning; Yang and Wong, ICCAD-94.
  — Based on max-flow min-cut theorem.

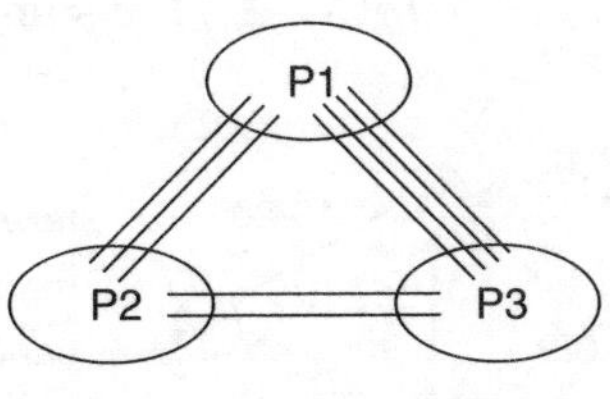

- Gate replication for partitioning : Yang and Wong, ICCAD-95.

- Performance-driven multiple-chip partitioning : Yang and Wong, FPGA'94, ED and TC-95.
- Multi-way partitioning with area and pin constraints : Liu and Wong, ISPD-97.
- Multi-resource partitioning : Liu, Zhu, and Wong, FPGA-98.
- Partitioning for time-multiplexed FPGAs : Liu and Wong, ICCAD-98.

8.16 FLOW NETWORKS

- A **flow network** $G = (V, E)$ is a **directed** graph in which each edge $(u, v) \in E$ has a **capacity** $c(u, v) > 0$.
- There is exactly one node with no incoming (outgoing) edges, called the **source** s (**sink** t).
- A **flow** $f : V \times V \longrightarrow R$ satisfies

  — **Capacity constraint :** $f(u, v) \leq c(u, v), \ \forall \ u, v \in V$.

  — **Skew symmetry :** $f(u, v) = -f(u, v), \ \forall \ u, v \in V$.

  — **Flow conservation :** $\Sigma_{v \in V} f(u, v) = 0, \ \forall \ u \in V - \{s, t\}$.
- The **value** of a flow $f : |f| = \Sigma_{v \in V} f(s, v) = \Sigma_{v \in V} f(v, t)$
- **Maximum-flow problem :** Given a flow network G with source s and sink t, find a flow of maximum value from s to t.

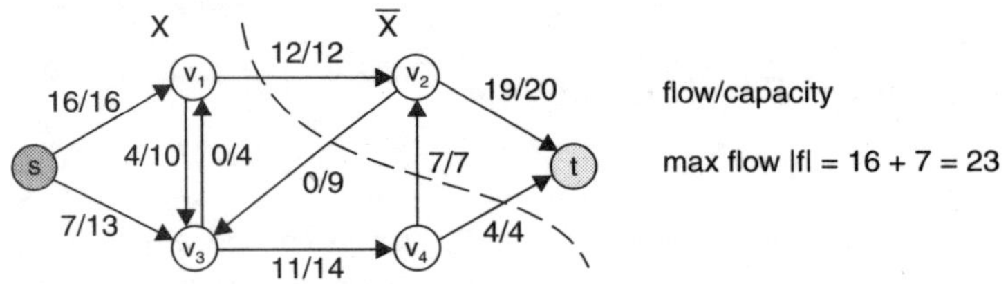

Max-Flow Min-Cut

- A **cut** $(X, \overline{X})$ of flow network $G = (V, E)$ is a partition of V into X and $\overline{X} = V - X$ such that $s \in X$ and $t \in \overline{X}$.

  — **Capacity of a cut :** $cap(X, \overline{X}) = \Sigma_{u \in X, v \in \overline{X}} c(u, v)$. (Count only **forward** edges !)
- **Max-flow min-cut theorem** Ford and Fulkerson, 1956.

  — f is a max-flow $\Leftrightarrow$ $|f| = cap(X, \overline{X})$ for some min-cut $(X, \overline{X})$.

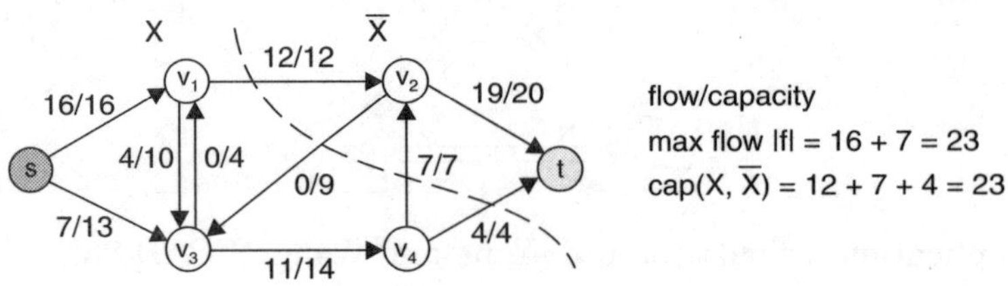

8.17 NETWORK FLOW ALGORITHMS

- An **augmenting path** p is a simple path from s to t with the following properties :
  — For every edge $(u, v) \in E$ on p in the **forward** direction (a **forward edge**), we have $f(u, \bar{v}) < c(u, v)$.
  — For every edge, $(u, v) \in E$ on p in the **reverse** direction (a **backward edge**), we have $f(u, v) > 0$.
- f is a max-flow $\Leftrightarrow$ no more augmenting path.
- First algorithm by Ford and Fulkerson in 1959 : $O(|E||f|)$; First **polynomial-time** algorithm by Edmonds and Karp in 1969 : $O(|E|^2 = |V|)$; Goldberg and Tarjan in 1985 : $O(|E||V| \lg (|V|^2/|E|))$, etc.

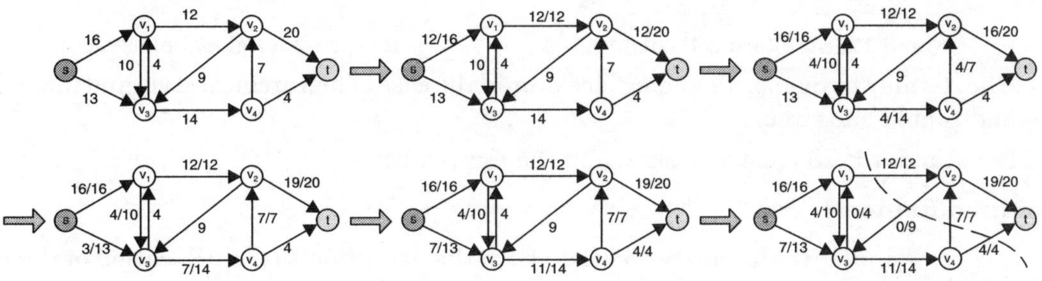

8.18 PLACEMENT

The standard cell placement is to map the ASIC components of cells onto the positions of the ASIC core area or the standard cell placing region, which is defined by rows.

The standard cells must be placed in the assigned region (ρ) such that the ASIC can be routed efficiently and the overall timing requirement can be satisfied. The placement has always been a key factor for achieving the physical designs with the optimised area usage, routing congestion and timing.

8.19 CLASSIFICATION OF PLACEMENT

The placement techniques are classified by:

1. Global placement   2. Detail placement

When the floorplan is first created, the standard cells are in a floating state. This means that they are placed arbitrarily in the ASIC core area.

Clusters. Refer to a group of standard cell that during placement are placed near to each other. The location of the cluster is undefined until all standard cells have been placed. This option is mainly used to control the closeness of the timing critical components during the placement.

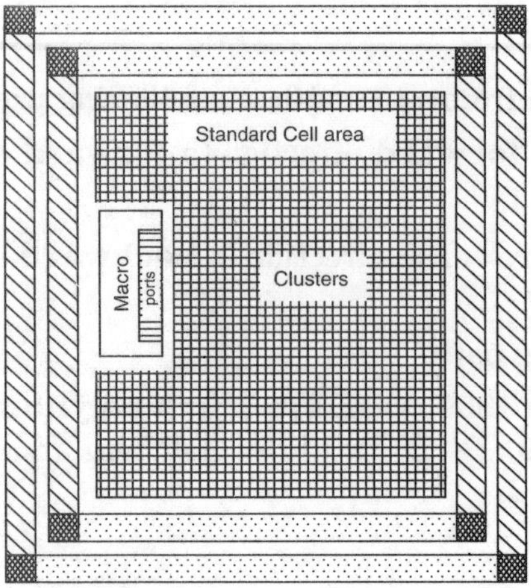

ETM

First macro ports are placed as many as required and then we place cluster.

Standard cell area

Fig. 8.12. Standard cell clusters.

<It is a type under global placement.>

Macros may be memories analog blocks or in the case of hierarchical style and individually placed and routed sub circuit.

The standard cell cluster is shown in the figure above.

Predefined Region:

It is very similar to the cluster method with the exception that the location of the region is defined prior to the standard cell placement. The way of this option is implemented is that 'a cluster or a group of standard cells' is created and then assigned to a particular area on the core of the ASIC.

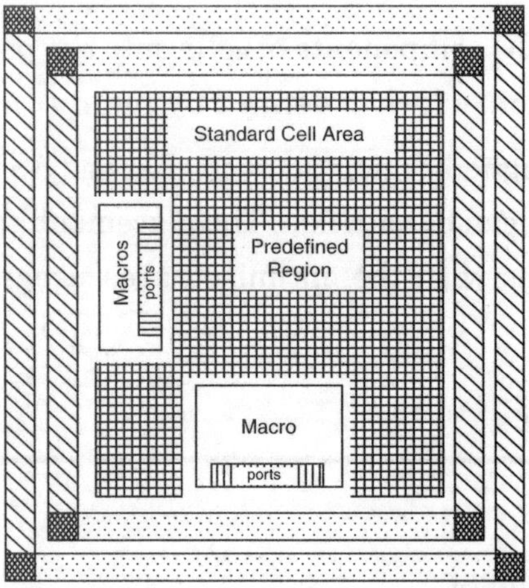

Note:

First predefined region is found and then selected (within it cluster is put) and then put as many macros as required.

Fig. 8.13. Predefined region

→ Regions can be soft or hard **soft region** is physical constraint where the logical module is assigned to a location in the core and boundary of the region.

→ Hard region is more rigorous than the soft region which defines a physical partition for the modular design. It has hard boundaries that prevent the standard cell crossing during placement.

8.20 GLOBAL PLACEMENT

There are 3 methods in global placement, used for partitioning during the global placement. The methods are also known as min-cut algorithms.

The min-cut algorithms are classified by

(*a*) Quadrature min-cut,

(*b*) Bisection min-cut,

(*c*) Slice and bisection min-cut.

8.20.1 Quadratic Min-Cut Algorithm

This algorithm partitions the ASIC core area into an equal number of instances in a vertical and horizontal direction and minimize the number of interconnections or cut-sizes in each direction. The vertical and horizontal cut lines start from the center of the core and recursively continue partitioning standard cell instances until the number of cut sizes are minimized and number more horizontal or vertical partitions are possible.

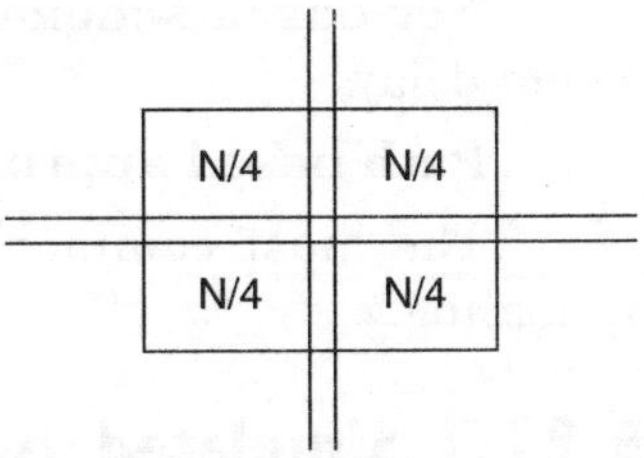

Advantages. It produces equilibrium between the horizontal and vertical routing without a congestion area. For this region most of the **place and route** tools utilizes this algorithm.

8.20.2 Bisection Min-Cut

This type of placement repeatedly divides or bisects the design by vertical or horizontal cut-lines depending on the standard cell row orientation until the slots contain one or two rows.

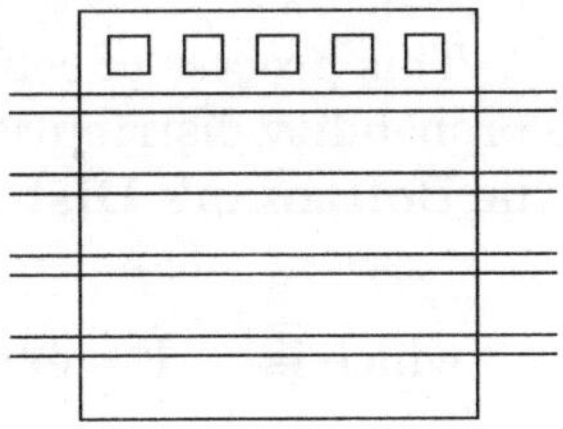

The position of each standard cell instances will be arbitrarily at this point.

Advantage. Block size area is minimum hence power consumption is low.

8.20.3 Slice and Bisection Method

This algorithm divides 'N' standard cells such that the smaller partition $\frac{N}{K}$ can be assigned to a standard cell row by horizontal slicing. The basic principle of the slice and bisection partitioning of *N* standard cells instances into *K*-no. of groups is $\frac{(K-1)N}{K} + \frac{N}{K} = N$

or $$\frac{K-1}{K} + \frac{1}{K} = 1$$

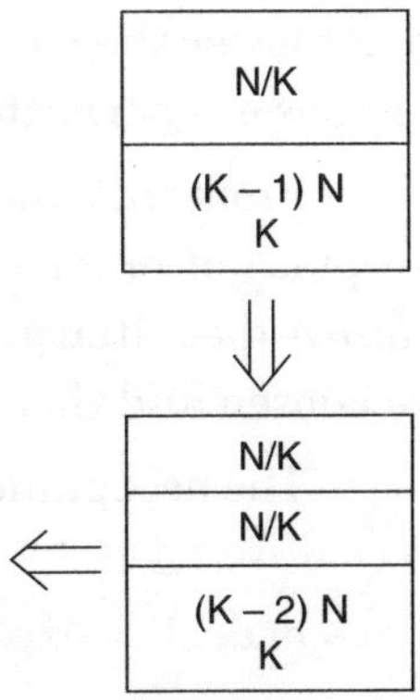

Slice and bisection

Advantage. Block size is further reduced and hence interconnection delay is further minimized. Cost of production is less. Power dissipation is minimized.

8.21 DETAIL PLACEMENT

A detail placement algorithm is executed to refine their placement based on congestion, timing and power requirements. **Congestion refinement** or **congestion driven placement** is more beneficial ASIC designs with very high density and the objective of the detail placer is to distance standard cell instances from each other such as more routing tracks are created among them.

Timing driven placement algorithm is classified as either

– **net based** or

– **path based** algorithm

Net based schemes try to control the delay on a signal path by imposing an upper bound delay.

Path based approaches apply the constraints to the delay paths of small sub-circuits.

The most commonly used algorithm used for the detail placement is simulated annealing.

8.21.1 Simulated Annealing

It is a simulation based placement technique. It is used as an iterative improvement algorithm during detail placement algorithm. The **objective** of this procedure is to find an optimal or near optimal placement for each pre-placed standard cell instances.

Theory. To produce a perfect crystal in a low energy state, material is heated to a high temperature and then slowly cooled down to form a solid.

This process of solidification is based on the assumption that at any given temperature the probability distribution of a change in structure is caused by energy level that is governed by the **Boltzman's Distribution Theorem.**

which is, $F = \exp\left(\frac{-E}{KT}\right)$

where, E = energy K = Boltzman's Constant

T = high temperature

When the temperature is high, the system can change radically and many changes that do not lower the energy level are allowed. As the temperature decreases the fewer non-optimal changes are permitted until an optimum state is achieved.

From this assumption, in the simulated annealing process ; a cost function, ΔC is used in the place of energy and the temperature, T is used as the control parameter to iteratively improve the initial pre-placed standard cell instances. The initial pre-placed standard cell instances are chosen and their placement is iteratively optimised.

The acceptance probability (P) = 1 when $\Delta C \leq 0$

and, $P = \exp\left(\frac{-\Delta C}{T}\right)$ when $\Delta C \leq 0$.

This algorithm begins with a very high temperature for the initial standard cell placement and the temperature slowly reduces according to an annealing schedule so that increases in cost have a progressively lower probability of acceptance.

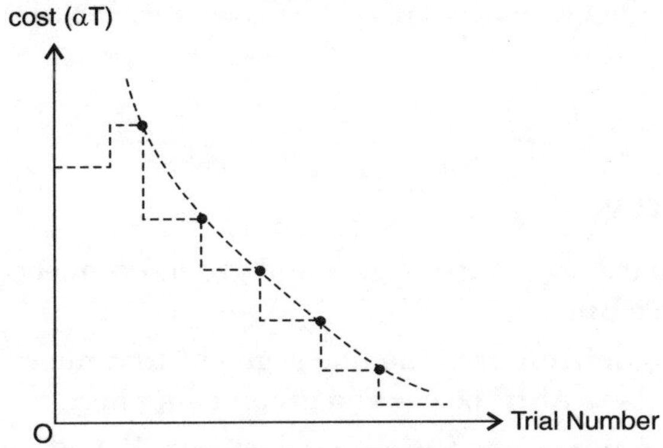

A cost function is used to indicate the desirability of a given placement and the changes in cost from the current placement to the new placement are computed.

These optimisation processes are iteratively repeated, all criteria are met and specified run-time are exceeded.

8.22 ROUTING

After the completion of standard cell placement, the next phase is to route the ASIC design and perform the extraction of routing.

The routing algorithms are mainly classified as

— **channel routing**

— **channel-less routing**

The most commonly used routing technique is the **channel routing.** The channel routers use the reserved space between the standard cell rows (routing channel) and feed-through dedicated routing areas inside the standard cell layout.

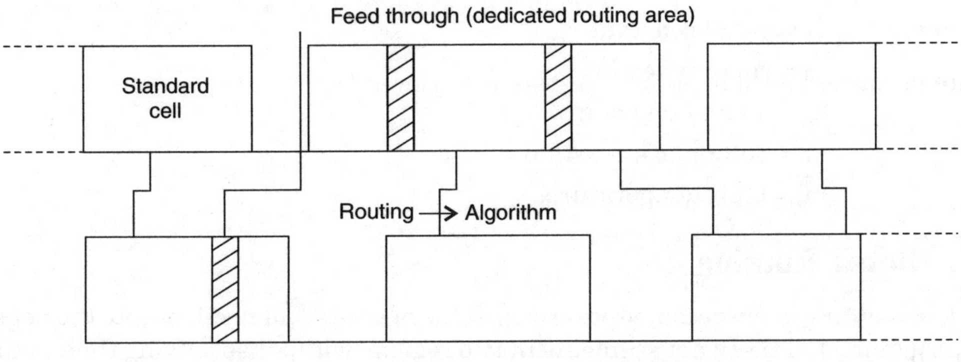

Fig. 8.14. Channel Routing Method

8.23 CLASSIFICATION OF ROUTING TECHNIQUES

Due to the inherent complexity ofASIC designs and very large number of interconnections associated with them the overall routing is performed in 3 stages.

1. Spatial Routing
2. Global Routing
3. Detail Routing.

8.23.1 Spatial Routing

It is used for standard cell, macro power and ground connections. Most spatial routers use the **line-probe algorithm.**

The **line-probe algorithm** uses the line segment to connect the standard cells macro power and the ground ports to ASIC power and ground supplies.

Line-probe routers use a generated connectivity list of sources and targets from connections.

The line segments are used to perform the routing according to the connectivity list, starting from the target and tracing the line segment until the source is reached.

These generated line segments do not pass through any obstructions. If they were to pass through an obstruction, then they might create an unfixable design rule violation.

In connecting the macro power and ground ports to the main power and ground nets, line port routers use the size of the ports to set the width of the power and the ground segments. For automatic width setting we need to create more power and ground ports to satisfy the current density requirements.

The excessive current density in an ASIC design can lead to an **Electro-migration problem** (if two metal interconnections pass through a field line, a flux is generated and that region dips a bit, ex-soldering) that reduces the Mean Time To Failure (MTTF) of the device. Most of the ASIC chips must have an MTTF of atleast 10 years. The MTTF is described by the Black's Equation.

Mathematically, $\text{MTTF} = \dfrac{A}{J^2} \exp.\left(\dfrac{E_a}{KT}\right)$ – Black's Eqn.

where, A = metal constant

J = current density

E_a = activation energy

k = Boltzman's constant

T = high temperature.

8.23.2 Global Routing

Global routing is the decomposition ofASIC design inter connections into the net segments and the assignment of these net segments in two regions without specifying their actual size of the layouts.

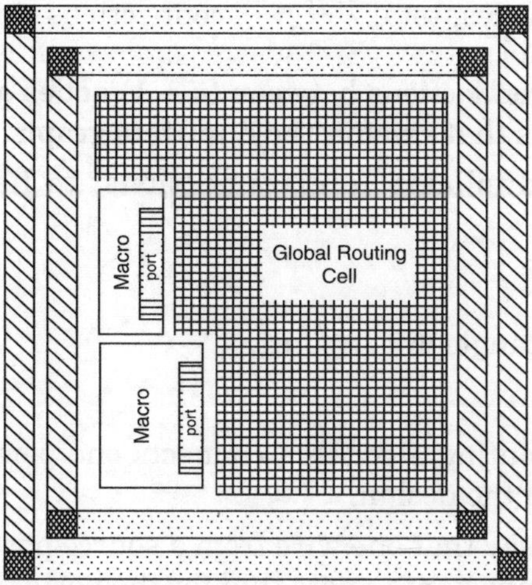

These routing regions are commonly known as Global Routing Cells, GRC. The density or capacity of these cells is defined as the maximum number of nets crossing the routing region and is a function of the number of routing layer, cell height in the vertical and horizontal direction, **minimum width** and the spacing of the wires as defined in the technology file.

Mathematically, $C_d = \dfrac{nh}{W + S}$

where, C_d = global routing cell density

n = number of routing layers that are available for vertical and horizontal direction.

h = height of the global routing cell length,

W = Minimum wire width.

S = Minimum wire spacing for the vertical and horizontal direction.

Depending on the global routing algorithm, there are several ways to estimate the associated wire length for each net in the design. They are :

(*a*) Complete Graph

(*b*) Source-to-sink

(*c*) Steiner Minimal Tree

(*d*) Minimum Spanning Tree

(*e*) Half Perimeter

(*f*) Minimum Chain.

8.23.2.1 *Complete Graph*

Complete graph has the connection from each port to every other port. In this method, all the interconnect length of the complete graph connections are added together and then divided by ***n*/2** where n is the number of ports.

i.e., to interconnect N nodes, total $\frac{n(n-1)}{2}$ connections exists.

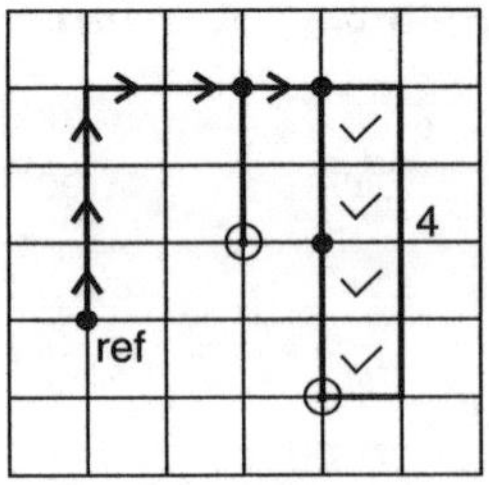

There are $(n-1)$ interconnections between $(n-1)$ nodes and there are some duplicate nodes that exist in the duplicate graph.

So we eliminate the duplicate interconnections, to have $\frac{n(n-1)}{2}$

CTM : Every small rectangle is 1 sq. unit.

8.23.2.2 *Source-to-sink*

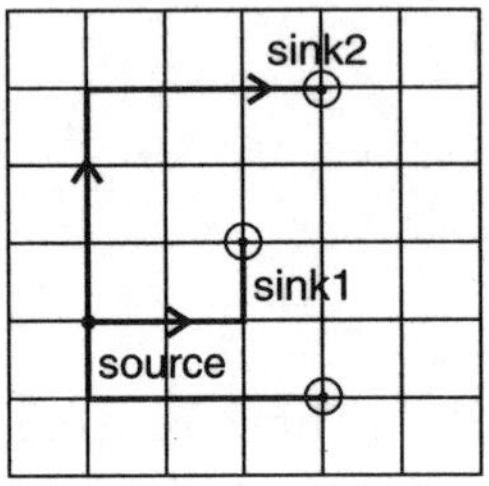

Source-to-sink global routing algorithm connects one port to all other ports for every net in the design.

In other words, we can say that one wire from a source port to other port is connected. It is important to note that for the ports that are located far from each other are connected directly from the source to the sink.

Disadvantage

This algorithm does not produce an accurate estimation.

Separate paths allocated for every source to sink path. No question of duplication. As a result not many sinks can be routed.

8.23.2.3 *Steiner Minimal Tree (SMT)*

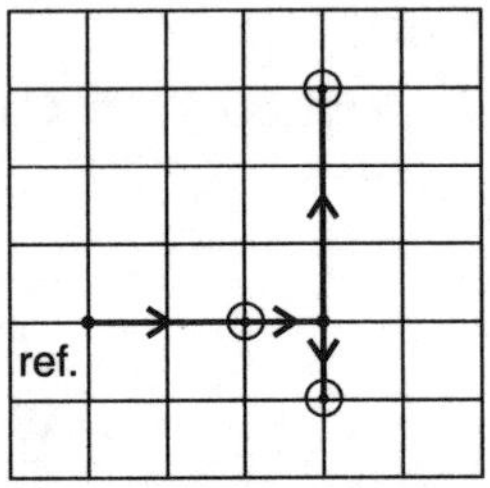

This type of algorithm is well suited for multiport nets. For a given set of ports in the design, the SMT connects these ports through some extra points called steiner points to achieve the minimum wire length. The most frequently used SMT is Rectilinear SMT which is the shortest interconnect using a rectangular grid.

8.23.2.4 *Minimum Spanning Tree*

In this procedure, all design ports and interconnections are considered to be a complete graph and the spanning tree is a subgraph that contains all the ports and interconnections.

The cost of the spanning tree is the sum of the lengths of each interconnection. Thus the different trees of the same interconnected network have different lengths.

The number of spanning trees is equal to 2, the power of number of ports.

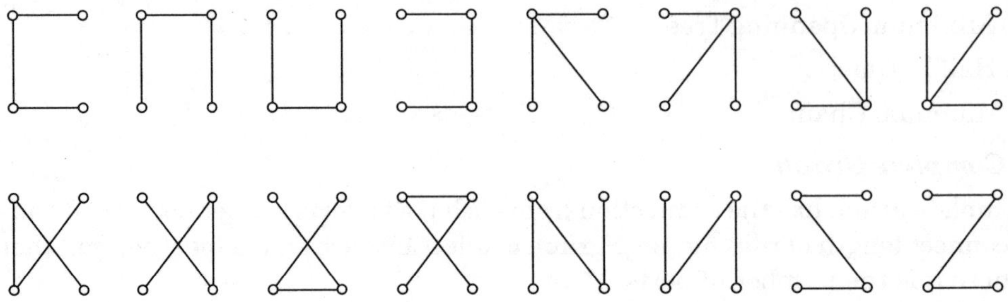

8.23.2.5 *Minimum Chain*

The minimum chain connectivity algorithm starts from one port and tries to connect that to the closest point and then to the next closest point in the chain sequence.

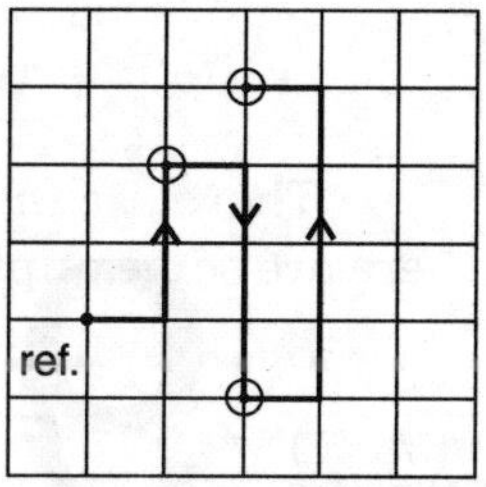

8.23.2.6 *Half Perimeter*

It uses a wiring bounding box that is the smallest-rectangle that encloses all ports. The half perimeter wire length estimation is one half of the wiring bounding box.

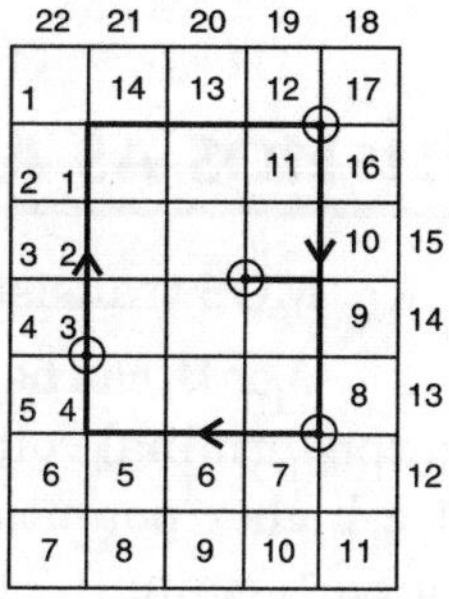

8.23.3 Detail Routing

Detail Routes use both horizontal and vertical routing grids for actual routing. The horizontal and vertical routing grids are defined in the technology file for all layers that are being used.

Detail Routers can be:

— Grid based

— Grid-less based

— Sub-grid based

(*a*) Grid based routers using the routing grid followers all routing segments.

(*b*) Grid-less based routers do not follow the routing grids explicitly but are dependent on the entire routing area and are not limited by the grids restriction.

(*c*) Sub-grid based routers follow the normal grid based routing technology but the sub-grid based routers consider these grids only as the guidelines.

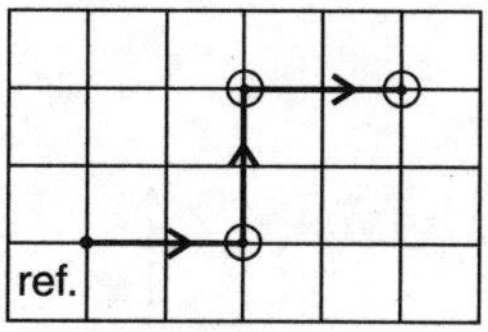

Grid based

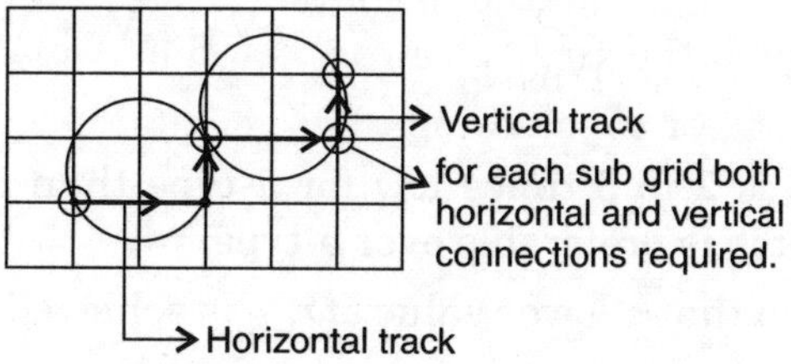

Sub-grid based

Chapter 9

ANALOG VLSI DESIGN

9.1 MOS AS A SWITCH

MOS transistor that is to be used as a switch.

A or B, can be the drain or the source of the MOS transistor depending on the terminal voltages (*e.g.*, for an *n*-channel transistor, if terminal A is at a higher potential than B, then terminal A is the drain and B is the source terminal). The 'ON' resistance consists of the series combination of r_0, r_S and whatever channel resistance exists.

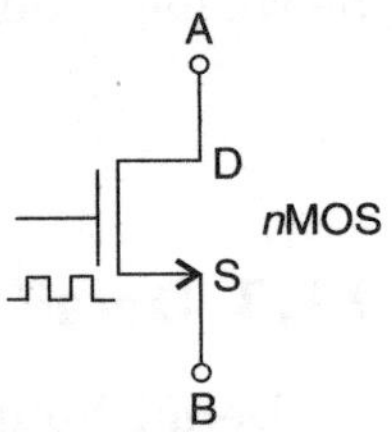

Fig. 9.1(*a*)

Typically by design, the contribution from r_D and r_S is small such that the primary consideration is the channel resistant.

In the 'ON' state of the switch the voltage across the switch should be small and V_{as} should be large. Therefore, the MOS device is assumed to be in the non-saturation region we know,

$$i_D = \frac{k'W}{L}\left[(V_{Gs} - V_T)\,V_{DS} - \frac{V_{DS}^{\,2}}{2}\right] \text{ where } 0 < V_{DS} < V_{Gs} - V_T$$

(V_{Gs} becomes V_{GD} if V_{DS} is negative).

The small signal channel resistance is given as,

$$r_{ON} = \left.\frac{1}{\dfrac{\delta I_D}{\delta V_{DS}}}\right|_Q = \frac{1}{\dfrac{k'W}{L}(V_{Gs} - V_T - V_{DS})}$$

☆ r_{ON} is 2 to 3 times low for *n*-type than *p*-type so, *n*-type switch is preferable over *p*-type.

It is seen that a lower value of r_{ON} is achieved for larger values of $\frac{W}{L}$.

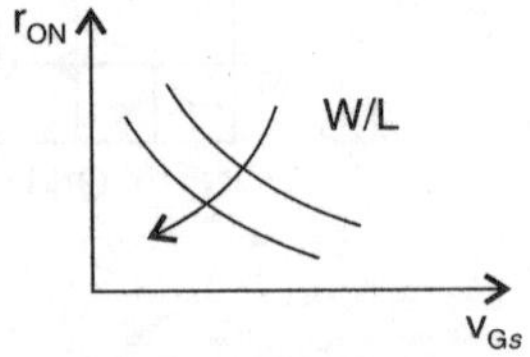

A plot of r_{ON} as a function of V_{Gs}.

Fig. 9.1(*b*)

When $V_{Gs} \approx V_T$ ($V_T = 0.7$ V) the r_{ON} approaches infinity, because the switch r_{ON} is turning 'off'.

For n-MOS, the gate voltage must be considerably larger than either the drain or source voltage in order to ensure that the MOS transistor is 'ON'.

Typically, the bulk is taken to the most negative potential for nMOS switch.

Consider the use of a switch to charge a capacitor is shown in Fig. 9.2.

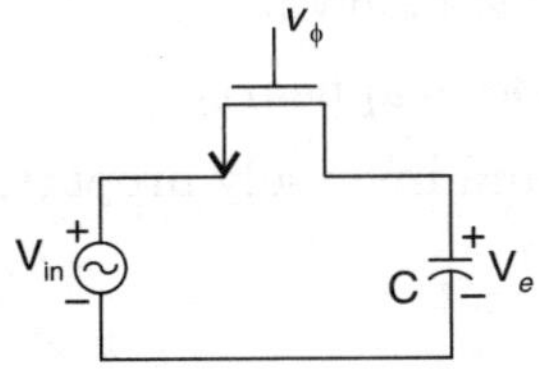

An application of MOS switch.

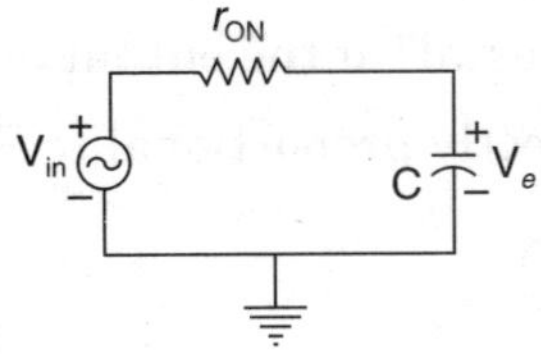

Model for the 'ON' state of the switch.

Fig. 9.2

An n channel transistor is used as a switch and V_ϕ is the control voltage (clk) applied to the gate.

The r_{ON} of the switch is important during the charge transfer phase of the circuit.

When V_ϕ goes high ($V_\phi > V_{in} + V_{rn}$) then M_1 connects C to the voltage source V_{in}. The equivalent circuit is given above in Figure.

It can be seen that C will charge in V_{in} with the time constant of r_{ON} C.

For successful operation,

$$r_{ON}\, C \leq T$$

where T = time and V_ϕ is high.

The worst case value for r_{ON} (the highest value), during the charging c, is when $V_D = 0$ and $V_{Gs} = V_\phi - V_{in}$.

It is derivable to keep the switch size as small as possible (minimize W × L) to minimize charge feedthrough from the gate.

One of the most serious limitations of monolithic switch is the clock feedthrough effect.

Clock feedthrough is happened by the coupling capacitance from the gate to both source and drain.

Channel Charge Injection

Assuming, $V_{in} \approx V_{out}$, then the total charge in the inversion layer as,

$$Q_{ch} = WLC_{ox}\,(V_{DD} - V_{in} - V_{th}).$$

where in saturation, $V_{Gs} \approx v_{DD}$ and L is the effective channel length.

When the switch turns 'off', then Q_{ch} exists through the source and drain terminals, this phenomenon is called "channel charge injection".

The charge injected to the left side of Fig. 9.3 is absorbed by the input source, without creating any error.

On the other side, the charge injected to the right side is deposited on C_H, introducing an error in the voltage stored on capacitors.

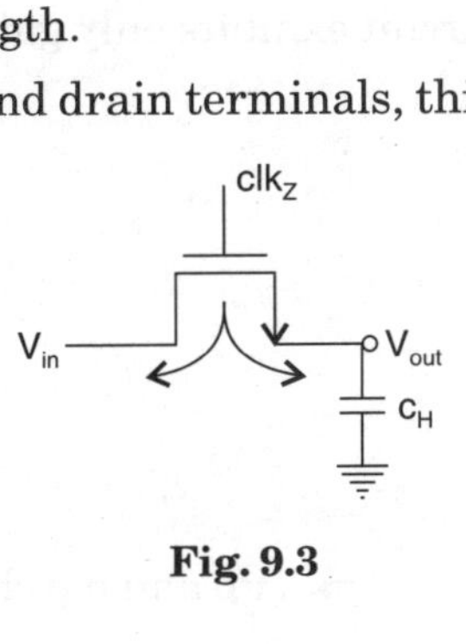

Fig. 9.3

For example, if half of Q_{ch} is injected onto C_H, then the resulting error equals,

$$\Delta v = \frac{WL\,C_{ox}\,(V_{DD} - V_{in} - V_{Th})}{2C_H}.$$

Another error for an nMOS switch is described below :

A negative "pedestal" at the output of a n-MOS is appeared.

The error is directly proportional to WLC_{ox} and inversely proportional to C_H.

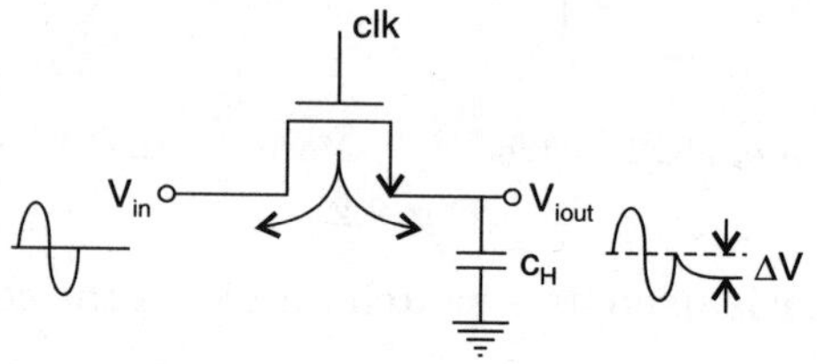

Fig. 9.4

Assume, all the charge is deposited on the capacitor then,

$$V_{out} = V_{in} - \frac{WL\,C_{ox}\,(V_{DD} - V_{in} - V_{Th})}{2C_H}$$

where phase shift between i/p and o/p is neglected.

$$= V_{in}\left(1 + \frac{WL\,C_{ox}}{2C_H}\right) - \frac{WL\,C_{ox}}{2C_H}\,(V_{DD} - V_{Th})$$

The o/p deviates from the ideal value through two effects :

(1) A non unity gain equal to $\left(1 + \frac{WL\,C_{ox}}{2C_H}\right)$

and (2) A constant offset voltage $\left\{-\frac{WL\,C_{ox}}{2C_H}\,(V_{DD} - V_{Th})\right\}$.

Since we have assumed channel charge is a linear function of the input voltage, the circuit exhibits only gain error and dc offset.

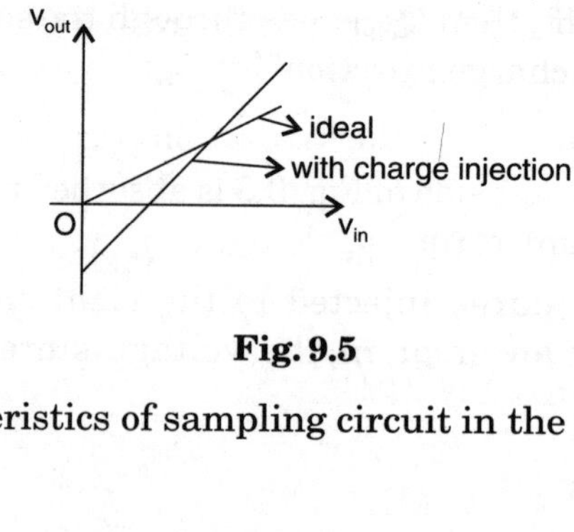

Fig. 9.5

⇒ i/p and o/p characteristics of sampling circuit in the presence of charge injection.

Clock Feedthrough

In addition to channel charge injection, MOS switch couples the clock transitions to the sampling capacitor through its gate-drain or gate-source overlap capacitance.

Assuming, the overlap capacitance is constant, then, error,

$$\Delta V = V_{ck} \cdot \frac{WL\,C_{ov}}{WC_{ov} + C_H} \quad ; C_{ov} = \text{overlap capacitance per unit width.}$$

The error ΔV is independent of the input level, manifesting itself as a constant offset in the i/p – o/p characteristics.

As with charge injection, clk feedthrough leads to a trade-off between speed and precission as well.

Clk feedthrough in a sampling ckt.

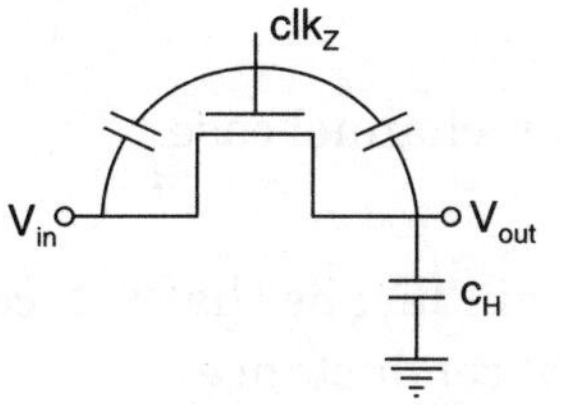

Fig. 9.6

Charge injection Cancellation :

(The advantage of *c*MOS switch over *n*MOS or *p*MOS switch).

This approach is used to lowering the effect of charge injection incorporates both *p*-MOS and *n*-MOS devices such that, the opposite charge packets injected by the two, canel each other.

For Δq_1 to cancel Δq_2, we must have,

$W_1L_1C_{ox}(V_{ck} - V_{in} - V_{Th}) = W_2L_2C_{ox}(V_{in} - V_{Th_p})$.

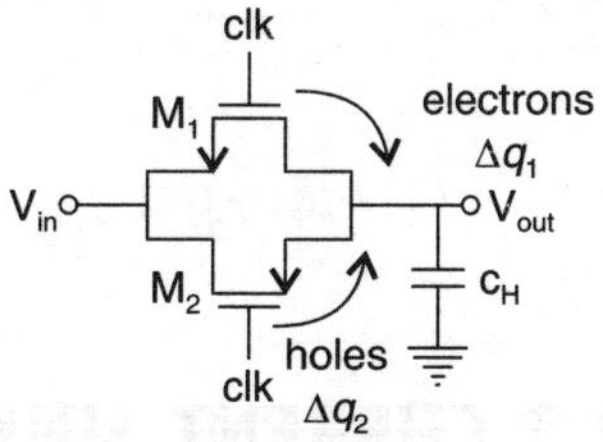

Fig. 9.7

9.2 MOS DIODE AND MOS RESISTOR

When the gate and drain of the MOS transistor are tied together, then the I-V characteristics are equally similar to a *p-n* junction diode, thus it is called MOS diode.

The large signal drain current in the saturation region,

$$I = I_D = \frac{k'}{2}\frac{W}{L}[(V_{as} - V_{Th})^2] = \frac{\beta}{2}(V_{as} - V_{Th})^2$$

or, $$V = V_{as} = V_{DS} = V_T + \sqrt{2I_D/\beta}$$

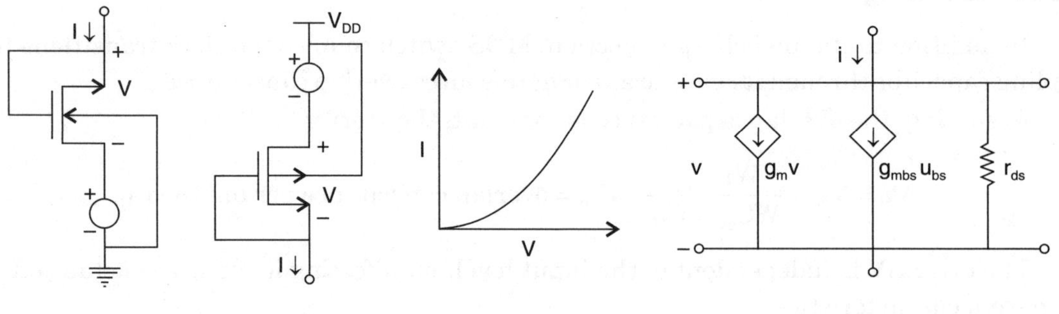

(a) (b) (c) (d)

Fig. 9.8

Fig. (*a*) *n*-channel active resistor.

(*b*) *p*-channel active resistor.

(*c*) I-V characteristics for *n*-channel case.

(*d*) Small-signal model.

Connecting the gate to the drain means that v_{DS} controls i_D and therefore, the channel transconductance becomes a channel conductance.

The small signal resistance of an MOS diode is

$$r_{out} = \frac{1}{g_m + g_{mbs} + g_{ds}} \approxeq \frac{1}{g_m}, \text{ where } g_m > g_{mbs} \text{ or } g_{ds}$$

When the transistor operated in the non-saturation region, the resistance can be calculated as,

$$r_{ds} = \frac{L}{KW\,(V_{as} - V_T)}.$$

9.3 CURRENT SINK AND CURRENT SOURCE

A current sink and a current source are two terminal components whose current at any instant of time is independent of the voltage across their terminals.

The current of a current sink or source flows from the positive node, through the sink or source, to the negative node.

A current sink typically has the negative node at V_{ss} and the current source has the positive node at V_{DD}.

In the current sink (Figure given below), the gate is taken to whatever voltage is necessary to create the desired value of current.

The point should be noted that in the non-saturation region the MOS device is not good current source.

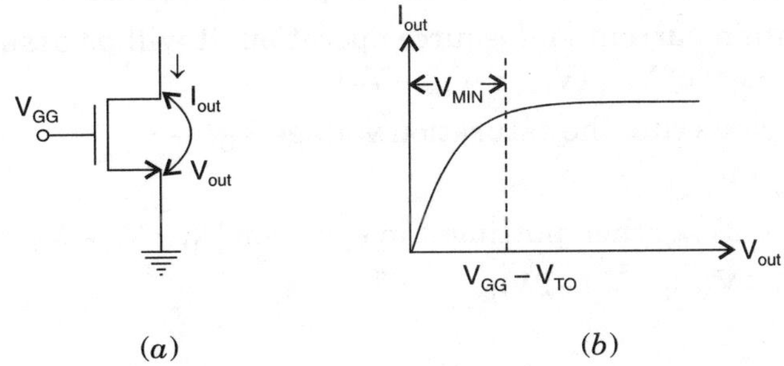

Fig. 9.9. (*a*) current sink (*b*) I-V characteristic of current sink.

The voltage across the current sink must be larger than V_{MIN} in order for the current sink to perform properly.

So, $V_{out} \geq V_{GG} - V_{TO}$.

If the source and bulk are both connected to ground; then the small-signal output resistance is given by,

$$r_{out} = \frac{1 + \lambda V_{DS}}{\lambda I_D} \approx \frac{1}{\lambda I_D}.$$

If the source and bulk are not connected to the same potential, the characteristics will not change as long as V_{BS} is a constant.

9.3.1 Current Source

The Fig. 9.10 given below, shows an implementation of a current source using a ϕ-channel transistor. The gate is taken to a constant potential as is the source.

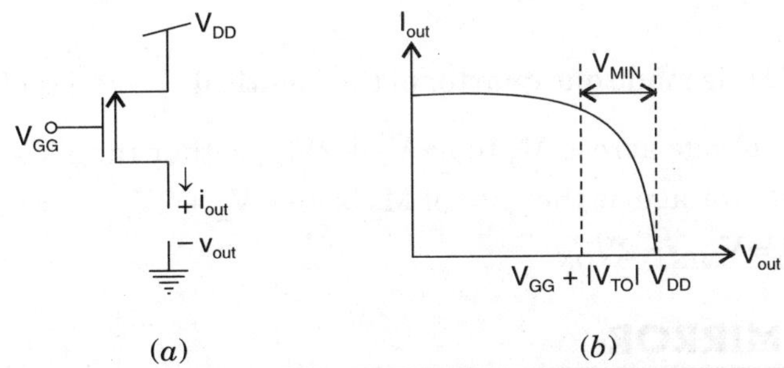

Fig. 9.10. (*a*) Current source (*b*) I-V characteristics of (*a*).

The source-drain voltage must be larger than V_{MIN} for this current source to work properly. This current source only works for values of V_{out} given by,

$$V_{out} \leq V_{GG} + |V_{TO}|.$$

9.3.2 The Standard Cascode Current Sink

If we ignore the bulk effects on M_2 and M_4 and assume that M_1, M_2, M_3 and M_4 are all matched with identical W/L ratios, then the gate-source voltage of each transistor can be expressed as $V_T + V_{ON}$.

At the gate of M_2, the voltage w.r.t. the lower power supply is $2V_T + 2V_{ON}$.

In order to maintain current sink/source operation, it will be assumed that M_1 and M_2 must have at least a voltage of V_{ON} ($V_{ON} = V_{Gs} - V_T$).

To find V_{MIN}, we can write, the saturation voltage V_D, as,

$$V_D \geq V_G - V_T.$$

Since, $V_{G2} = 2V_T + 2V_{ON}$, then putting it in equation $V_D \geq V_G - V_T$, we get,

$$V_{D_2\,(min)} = V_{MIN} = V_T + 2V_{ON}.$$

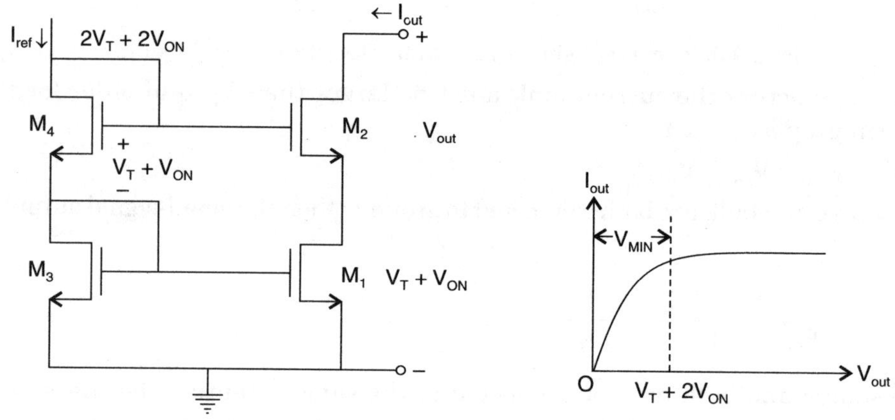

Fig. 9.11

V_{MIN} is dropped across both M_1 and M_2. The drop across M_2 is V_{ON} while the drop across M_1 is $V_T + V_{ON}$.

The equation $V_{DS\,(sat)} = V_{as} - V_T = V_{ON}$ and V_{MIN} of figure above, could be reduced by V_T and still keep both M_1 and M_2 is saturation.

The W/L ratio of M_4 is made one quarter of the identical $\frac{W}{L}$ ratios of M_1 through M_3. This causes the gate source voltage across M_4 to be $V_T + 2V_{ON}$ rather than $V_T + V_{ON}$.

Consequently, the voltage at the gate of M_2 is now $V_T + 2V_{ON}$.

Now, $\quad V_{D_2(min)} = V_{MIN} = 2V_{ON}$.

9.4 CURRENT MIRROR

The basic current mirror shows in figure below.

A current flows through M_1 corresponding to V_{Gs_1}. Since $V_{Gs_1} = V_{Gs_2}$ ideally the same current, or a multiple of the current in M_1. Flows through M_2. If the MOSFETs are the same size. The same drain current flows in each MOSFET, provides M_2 stays in the saturation region.

The current I_{D_1} is given by,

$$I_{D_1} = \frac{\beta_1}{2}\,(V_{Gs_1} - V_{THN})^2 \qquad ...(1)$$

While the o/p current, assuming M_2 is in saturation, flowing in M_2 is,

$$I_{D_2} = I_0 = \frac{\beta_2}{2}(V_{Gs_2} - V_{THN})^2 \quad ...(2)$$

Since, $V_{Gs_1} = V_{Gs_2}$, the ratio of the drain currents is given by,

$$\frac{I_{D_2}}{I_{D_1}} = \frac{W_2/L_2}{W_1/L_1} = \frac{W_2L_1}{W_1L_2} = \frac{\beta_2}{\beta_1} \quad ...(3)$$

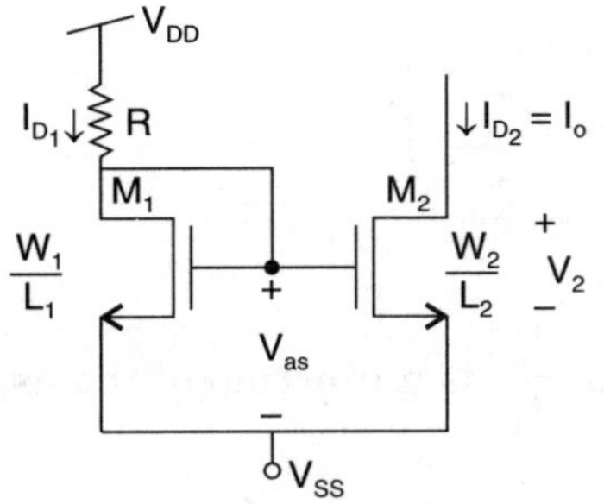

Fig. 9.12

This equation shows how to adjust the W/L ratio of the two devices to achieve the desired o/p current, I_{D_2}. This equation, however, does not show how the o/p current will change with the voltage across M_2, V_0.

The ref. drain current, $I_{ref} = I_{D_1} = \dfrac{V_{DD} - V_{as} - V_{ss}}{R} = \dfrac{KP.W_1}{2L_1}(V_{Gs1} - V_{THN})^2$ ...(4)

This minimum voltage, V_{min}, across the current sink is set by the requirement that M_2 remain in saturation, that is, $V_{min} = V_{DS,\,sat} = V_{Gs} - V_{THN}$.

The o/p resistance of the current source is simply the o/p resistance of M_2, or,

$$r_{O2} = \frac{1}{\lambda I_0} = \frac{1}{\lambda I_{D_2}}.$$

When, designing CMOS current mirrors, the values for V_{Gs} and L are selected before solving for W to get the desired current.

$$\frac{I_{D_2}}{I_{D_1}} = \frac{W_2}{W_1}.$$

The current mirror uses the principle that, if the gate-source potentials of two identical MOS transistors are equal, the channel currents should be equal.

A simple n-channel current mirror is implemented in the circuit, given below.

M_1 is in saturation region because $V_{DS_1} = V_{as_1}$.

For the adjoining gates of the M_1 and M_2 are connected, together then the V_{Gs} of the both of the transistor is same.

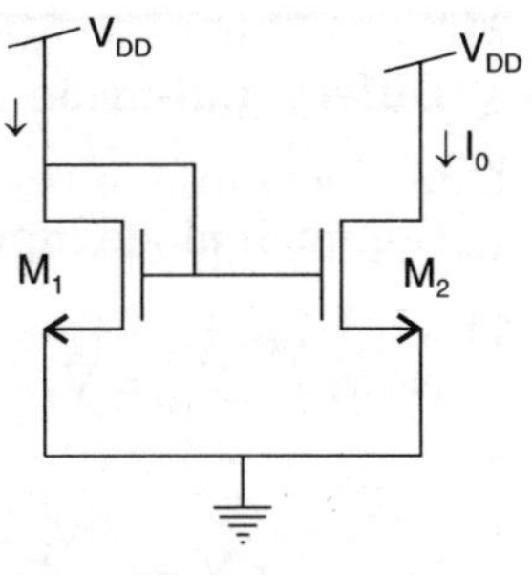

Fig. 9.13

So, M_2 is also acts in saturation region.

$$I_L = K_1' \frac{W_1}{L_1} (V_{as} - V_{T_1})^2 \times (1 + \lambda V_{DS1})$$

$$I_0 = K_2' \frac{W_2}{L_2} (V_{as} - V_{T_2})^2 \times (1 + \lambda V_{DS2}).$$

If, $K_1' = K_2'$ and $V_{T_1} = V_{T_2} = V_T$ then,

$$\frac{I_1}{I_0} = \left(\frac{W_1 L_2}{W_2 L_1}\right)\left(\frac{1 + \lambda V_{DS_1}}{1 + \lambda V_{DS_2}}\right).$$

If, $V_{DS_2} = V_{DS_1}$, then,

$$\frac{I_0}{I_1} = \left(\frac{W_2 L_1}{W_1 L_2}\right); \text{ So, } \frac{I_0}{I_1} \text{ is a function of the aspect ratio.}$$

If, $\frac{W_1}{L_1} = \frac{W_2}{L_2}$ then, $I_0 = I_1$.

⇒ Circuit is called current mirror.

If, $\frac{W_1}{L_1} = \frac{W_2/L_2}{A}$ then, $I_0 = AI_1$.

⇒ Current amplifier.

*[Small amount of amplification will possible because the aspect ratio should not very high].

If $V_{DS_1} \neq V_{DS_2}$ then,

$$\frac{I_0}{I_1} = \frac{1 + \lambda V_{DS_2}}{1 + \lambda V_{DS_1}}$$

⇒ The cascode current transistor circuit.

9.5 DIFFERENTIAL AMPLIFIER

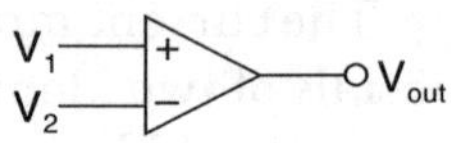

Differential-mode input voltage = V_{ID}. Common mode input voltage = V_{IC}.

Common mode input voltage = V_{IC}.

Now, $V_{ID} = V_1 - V_2$ $\qquad V_1 = V_{IC} + \frac{V_{ID}}{2}$

$$V_{IC} = \frac{V_1 + V_2}{2} \qquad V_2 = V_{IC} - \frac{V_{ID}}{2}.$$

$$V_{out} = A_{VD} V_{ID} \pm A_{VC} V_{IC}$$

$$= A_{VD} (V_1 - V_2) \pm A_{VC} \left(\frac{V_1 + V_2}{2}\right).$$

where,

A_{VD} = Differential-mode voltage gain.

A_{VC} = Common-mode voltage gain.

The differential amplifier using nMOS is given below :

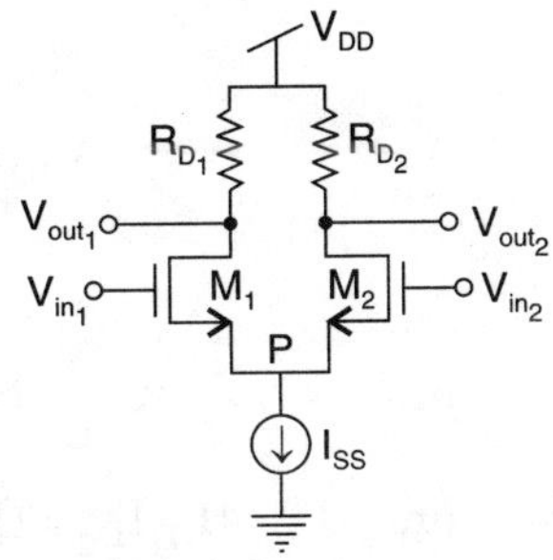

where, $R_{D_1} = R_{D_2} = R_D$.

From the fig. we can say that, $v_{in_1} - v_{in_2}$ varies from $-\alpha$ to α.

The transfer characteristics (Input-output characteristics) of a differential pair is given below :

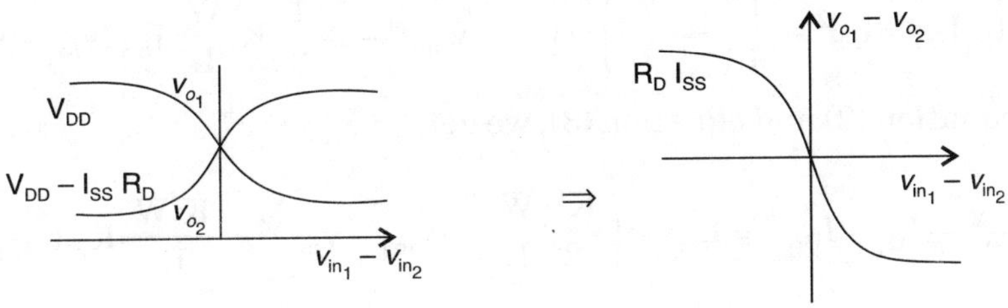

If $v_{in_1} << v_{in_2} \rightarrow M_1$ is off, M_2 is ON, $I_{D_2} = I_{SS}$, then $v_{out_1} = v_{DD}$ and $v_{out_2} = v_{DD} - R_D I_{SS}$.

As, $v_{in_1} \approx v_{in_2}$, then M_1 gradually turns 'ON', drawing a fraction of I_{SS} from R_{D_1} and hence lowering v_{out_1}.

Since, $I_{D_1} + I_{D_2} = I_{SS}$, the drain current of M_2 decreases and v_{out_2} increases.

For the differential pair,

$$v_{out_1} = v_{DD} - I_{D_1} R_{D_1}$$

$$v_{out_2} = v_{DD} - I_{D_2} R_{D_2}$$

$\therefore \quad v_{out_1} - v_{out_2} = I_{D_2} R_{D_2} - I_{D_1} R_{D_1} = (I_{D_1} - I_{D_2}) R_D$ if, $R_{D_1} = R_{D_2} = R_D$.

Considered, M_1 and M_2 are saturated and $\lambda = 0$.

Since, the voltage at node P is equal to $v_{in_1} - v_{Gs_1}$ and $v_{in_2} - v_{Gs_2}$,

$\therefore \quad (v_{in_1} - v_{in_2}) = (v_{gs1} - v_{gs2})$

For a square law device, we have,

$$(v_{gs} - v_{Th})^2 = \frac{I_D}{\frac{K'}{2} \cdot \frac{W}{L}}$$

and therefore,

$$v_{gs} = \sqrt{\frac{2I_D}{K' \frac{W}{L}}} + v_{Th}.$$

$$\therefore \quad v_{in_1} - v_{in_2} = (v_{gs_1} = v_{gs_2}) = \sqrt{\frac{2I_{D_1}}{K'\frac{W}{L}}} - \sqrt{\frac{2I_{D_2}}{K'\frac{W}{L}}} \qquad (\text{when, } V_{T_1} = V_{T_2} = V_T)$$

$$\text{or,} \quad (v_{in_1} - v_{in_2})^2 = \frac{2}{K'\frac{W}{L}}(I_{SS} - 2\sqrt{I_{D_1}I_{D_2}}) \qquad (\because \ I_{D_1} + I_{D_2} = I_{SS})$$

$$\text{or,} \quad \frac{1}{2}K'\frac{W}{L}(v_{in_1} - v_{in_2})^2 - I_{SS} = -2\sqrt{I_{D_1}I_{D_2}} \qquad \text{...(1)}$$

$$\text{or,} \quad 4I_{D_1}I_{D_2} = (I_{D_1} + I_{D_2})^2 - (I_{D_1} - I_{D_2})^2$$

$$\text{or,} \quad 4I_{D_1}I_{D_2} = I_{SS}^2 - (I_{D_1} - I_{D_2})^2 \qquad \text{...(2)}$$

Again, from equation (1),

$$4I_{D_1}I_{D_2} = I_{SS}^2 + \frac{1}{4}\left(K'\frac{W}{L}\right)^2 (v_{in_1} - v_{in_2})^4 - 2\cdot\frac{1}{2}K'\frac{W}{L} I_{SS}(v_{in_1} - v_{in_2})^2 \qquad \text{...(3)}$$

Equating equation (2) and equation (3), we get,

$$I_{SS}^2 - (I_{D_1} - I_{D_2})^2 = I_{SS}^2 + \left(\frac{K'}{2}\frac{W}{L}\right)^2 (v_{in_1} - v_{in_2})^4 - \frac{K'W}{L} I_{SS}(v_{in_1} - v_{in_2})$$

$$\text{or,} \quad (I_{D_1} + I_{D_2})^2 = -\frac{1}{4}\left(K'\frac{W}{L}\right)^2 (v_{in_1} - v_{in_2})^4 + I_{SS}\left(\frac{K'W}{L}\right)(v_{in_1} - v_{in_2})^2.$$

$$\text{Thus,} \quad (I_{D_1} + I_{D_2})^2 = \frac{1}{2}\mu_n c_{ox}\frac{W}{L}(v_{in_1} - v_{in_2})\sqrt{\frac{4I_{SS}}{\mu_n c_{ox}\frac{W}{L}} - (v_{in_1} - v_{in_2})^2}$$

where $K' = \mu_n c_{ox}$.

If, $v_{in_1} = v_{in_2}$ then $I_{D_1} - I_{D_2} = 0$ or $I_{D_1} = I_{D_2}$.

As $|v_{in_1} - v_{in_2}|$ increases from zero, then $|I_{D_1} - I_{D_2}|$ also increases.

$$* \therefore \quad G_m = \frac{\delta(\Delta I_D)}{\delta(\Delta V_{in})} = \frac{1}{2}\mu_n c_{ox}\frac{W}{L}\frac{\frac{4I_{SS}}{\mu_n c_{ox} W/L} - 2\Delta V_{in}^2}{\sqrt{\frac{4I_{SS}}{\mu_n c_{ox} W/L} - \Delta V_{in}^2}}$$

For, $\Delta V_{in} = 0$, $\quad G_m = \sqrt{\mu_n c_{ox}\frac{W}{L} I_{SS}}$

Since, $V_{out_1} - V_{out_2} = \Delta I \cdot R_D = R_D G_m \cdot \Delta v_{in}$

$$\therefore \quad \boxed{A_v = \sqrt{\mu_n c_{ox}\frac{W}{L} I_{SS}} \cdot R_D.}$$

The variations drain currents and overall transconductance of a differential pair, versus input voltage is shown below :

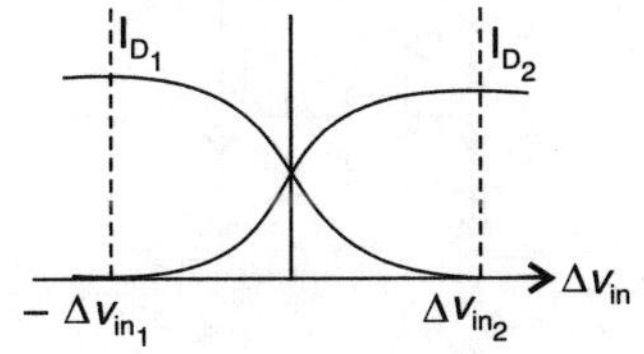

Now, the resistors are replaced by the two *p*MOS transistors of a differential pair amplifier. The circuit is given below :

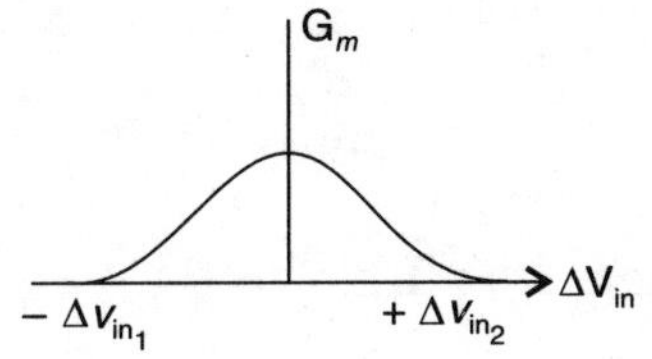

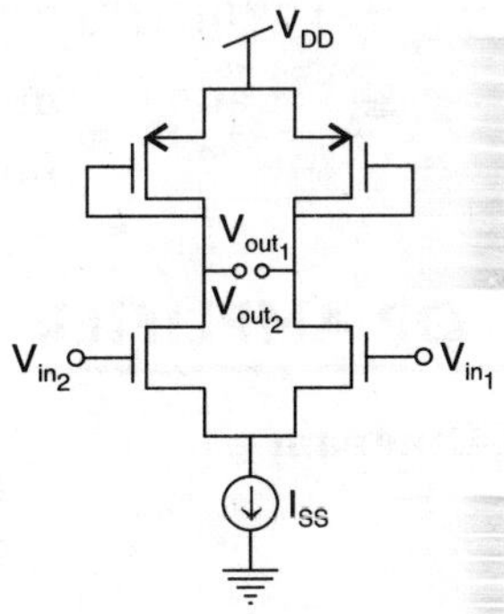

$$A_v = -g_{mN}\left(\frac{1}{g_{mp}} \parallel r_{ON} \parallel r_{OP}\right) \approx \frac{g_{mN}}{g_{mP}}$$

$$\therefore \quad A_v = -\sqrt{\frac{\mu_n (W/L)_N}{\mu_p (W/L)_P}}$$

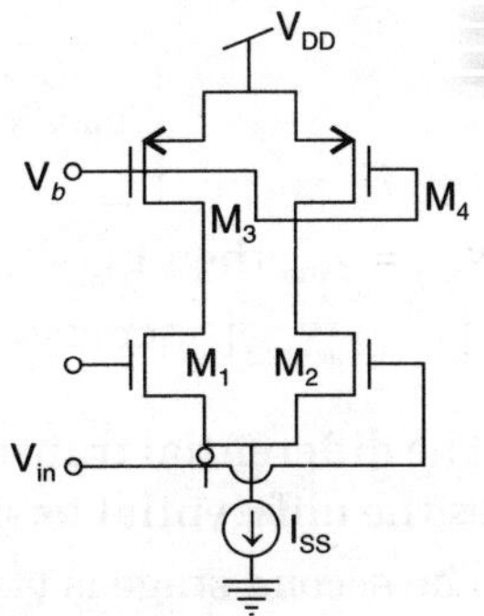

$$* A_v = -g_{mN}(r_{ON} \parallel r_{OP}).$$

9.5.1 Ideal differential Amp. which is used in OP-Amp

$I_{D_1} = I_{D_3}$ | $\because$ M_1 and M_3 are in saturation.

$I_{D_3} = I_{D_4}$ since M_3 and M_4 are used in current mirror configuration.

$\therefore \quad I_{D_1} = I_{D_4} = I_{D_2}$ for $V_{in_1} = V_{in_2}$.

$\therefore \quad (V_{out_1} - V_{out_2}) \propto (V_{in_1} - V_{in_2}) \propto (I_{D_1} - I_{D_2})$.

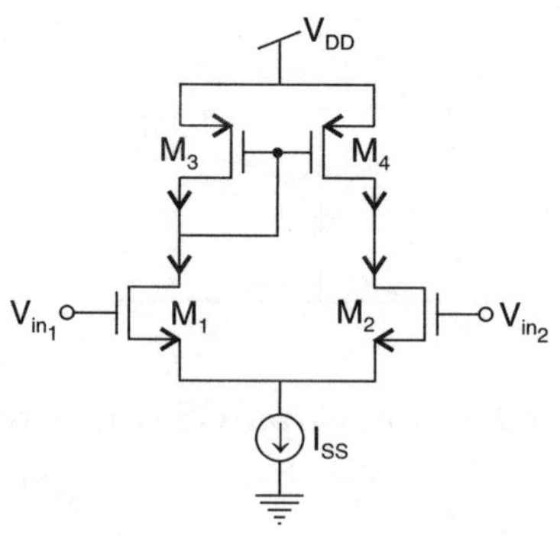

$\therefore \quad I_{out} = I_{D4} - I_{D2}\ ;\ I_{D2} \neq I_{D4}.$

$\qquad = 0 \text{ for } I_{D2} = I_{D4}.$

If, $V_{in1} > V_{in2}$, $I_{D1} > I_{D2}$. Then, $I_{D4} > I_{D2}$.

$\therefore \quad I_{out} = I_{D4} - I_{D2} > 0.$

$$A_v = G_m R_L = \frac{dv_{out}}{dv_{ID}}.$$

9.6 OP-AMPLIFIER USING MOS

Block Diagram

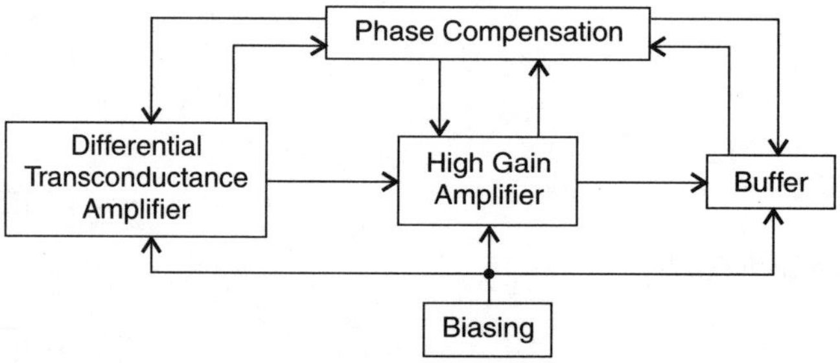

The differential transconductance stage forms the inputs of the OP-Amp. and sometimes provides the differential to single ended conversion.

The second stage is typically an inverter. If the differential-input stage does not perform the differential-to-single ended conversion, then it is accomplished in the second stage inverter.

If the OP-Amp drives a low resistance load, then the 2nd stage must be followed by a buffer stage whose objective is to lower the o/p resistance and maintain a large signal swing.

Bias circuits are provided to establish the proper operating point for each transistor in its quescent stage.

Design of the two-stage OP-Amp :

Assuming, $\quad g_{m_1} = g_{m_2} = g_{m_I}$

$\qquad g_{m_6} = g_{m_{II}}.$

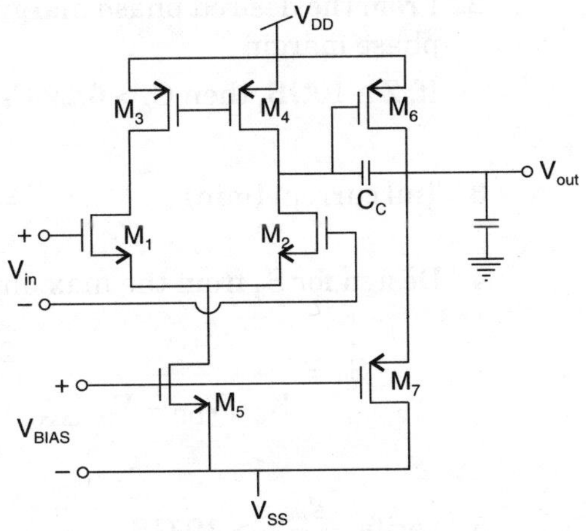

$$\therefore \quad g_{ds_2} + g_{ds_4} = G_I$$

$$g_{ds_6} + g_{ds_7} = G_{II}$$

Slow rate :

$$SR = \frac{I_5}{C_c}$$

1st stage gain :

$$A_{v1} = -\frac{g_{m_1}}{g_{ds_2} + g_{ds_4}} = \frac{-2g_{m_1}}{I_5(\lambda_2 + \lambda_4)}$$

Second stage gain :

$$A_{v2} = -\frac{g_{m_6}}{g_{ds_6} + g_{ds_7}} = -\frac{g_{m_6}}{I_6(\lambda_6 + \lambda_7)}.$$

Gain band width,

$$G_B = \frac{g_{m_1}}{C_c}$$

Output pole, $p_2 = -\dfrac{g_{m_6}}{C_L}$

RHP zero : $Z_1 = \dfrac{g_{m_6}}{C_c}$.

Positive CMR $V_{in\,(max)} = V_{DD} - \sqrt{\dfrac{I_5}{\beta_3}} - |V_{TO_3}|_{(max)} + V_{T1(min)}$

Negative CMR $V_{in\,(min)} = V_{SS} + \sqrt{\dfrac{I_5}{\beta_1}} + V_{T1\,(max)} + V_{DS5(sat)}$

Saturation voltage,

$$V_{DS\,(sat)} = \sqrt{\frac{2I_{DS}}{\beta}}.$$

Un-buffered Op-Amp. Design :

1. Choose the smallest device length that will keep the channel modulation parameter constant and give good matching for current-mirrors.

2. From the desired phase margin, choose the minimum value for C_L : that is, for a 60° phase margin :

   If, $Z \geq 10GB$, then $C_c > 0.22\, C_L$.

3. Tail current (min) $\Rightarrow I_5 = (S.R.) \cdot C_c \approx 10 \left(\dfrac{V_{DD} + |V_{SS}|}{2T_S} \right)$.

4. Design for S_3 from the max. input voltage specification,

$$S_3 = \frac{2I_3}{K_3' \,[V_{DD} - V_{in\,(max)} - |V_{TO_3}|_{max} + V_{T_1\,(min)}]^2} \geq 1$$

5. Verify, $\dfrac{g_{m_3}}{2cg_{s_3}} > 10$ GB.

6. $g_{m_1} = (G.B.) \cdot C_c$

$$\Rightarrow \quad S_1 = S_2 = \frac{g_{m_2}}{K' I_5}.$$

7. $V_{DS\,(sat)} = V_{in\,(min)} - V_{SS} - \sqrt{\dfrac{I_5}{\beta_1}} - V_{T_1\,(max)} \geq 100\, mV.$

$$S_5 = \frac{2I_5}{K_5' \,[v_{DSS(sat)}]^2}.$$

8. $g_{m_6} = 2.2\, g_{m_2} (C_L/C_c)$.

   Let, $V_{SG_4} = V_{SG_6}$ then, $S_6 = S_4 \cdot \dfrac{g_{m_6}}{g_{m_4}}$

$$\therefore \quad I_6 = \frac{g_{m_6}^2}{2K_6' \cdot S_6}.$$

9. $S_6 = \dfrac{g_{m_6}}{K_6' \, V_{DS_6\,(sat)}}$.

10. $S_7 = \left(\dfrac{I_6}{I_5} \right) S_5$.

11. $A_v = \dfrac{2 g_{m_2} g_{m_6}}{I_5\, (\lambda_2 + \lambda_3)(\lambda_6 + \lambda_7)}$.

    $P_{diss} = (I_5 + I_6)(V_{DD} + |V_{SS}|)$.

9.7 SOURCE FOLLOWER

The source-follower (common-drain stage) can operate as a voltage buffer.

The source follower senses the signal at the gate and drives the load at the source allowing the source potential to "follow" the gate voltage.

The basic source follower circuit is shown in figure given below:

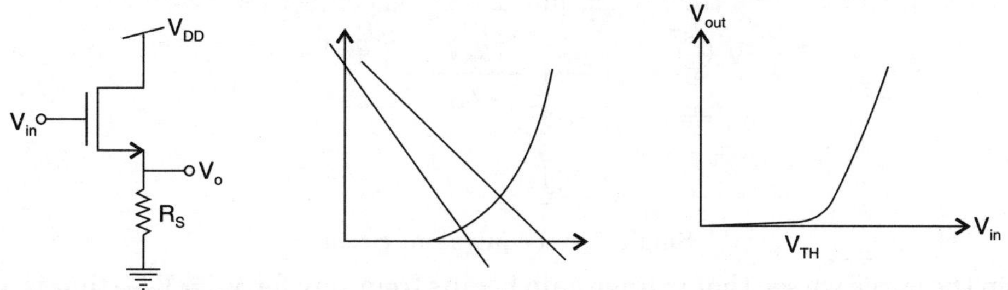

By analyzing the large signal behaviour, when $V_{in} < V_{TH}$, then M_1 is "off" and $V_{out} = 0$.

As $V_{in} > V_{TH}$, M_1 turns 'on' in saturation (for typical values of V_{DD}) and I_{D_1} flows through R_s.

As $V_{in} >> V_{TH}$, V_{out} follows the input with a difference (level shift) equal to V_{as}.

We can express the input-output characteristics as :

$$\frac{1}{2}\mu_n c_{ox}\frac{W}{L}(V_{in} - V_{TH} - V_{out})^2 R_s = V_{out}.$$

Now, differentiating both sides w.r.t. V_{in}, of equation and we will get the small-signal gain :

$$\frac{1}{2}\mu_n c_{ox}\frac{W}{L}\,2(V_{in} - V_{TH} - V_{out})\left(1 - \frac{\delta V_{TH}}{\delta V_{in}} - \frac{\delta V_{out}}{\delta V_{in}}\right)R_s = \frac{\delta V_{out}}{\delta V_{in}}.$$

Since, $$\frac{\delta V_{out}}{\delta V_{in}} = \frac{1}{\eta}\frac{\delta V_{TH}}{\delta V_{in}}.$$

$$\therefore \quad \frac{\delta V_{out}}{\delta V_{in}} = \frac{\mu_n c_{ox}\frac{W}{L}(V_{in} - V_{TH} - V_{out})R_s}{1 + \mu_n c_{ox}\frac{W}{L}(V_{in} - V_{TH} - V_{out})R_s(1+\eta)}.$$

Again, $g_m = \mu_n c_{ox}\frac{W}{L}(V_{in} - V_{TH} - V_{out})$.

$$\therefore \quad A_v = \frac{g_m R_s}{1 + (g_m + g_{mb})R_s}.$$

From the small-signal equivalent circuit we get, $V_{in} - V_1 = V_{out}$.

$$V_{in} - V_1 = V_{out}.$$

$$V_{bs} = -V_{out}.$$

$$\therefore \quad g_m V_1 - g_{mb} V_{out} = \frac{V_{out}}{R_s}$$

$$\frac{V_{out}}{V_{in}} = \frac{g_m R_s}{1 + (g_m + g_{mb}) R_s}.$$

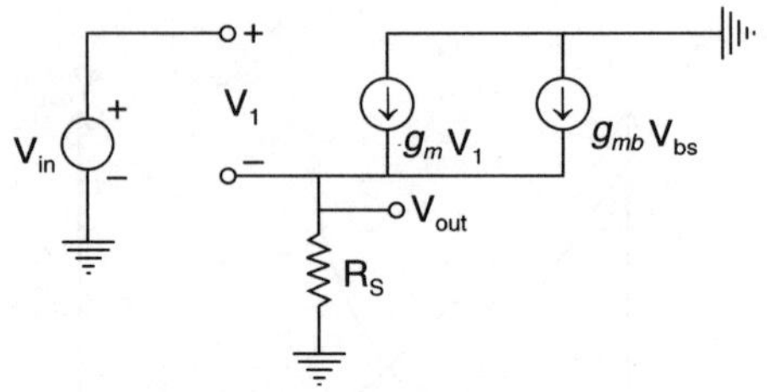

Small signal equivalent circuit

From the curve we see that voltage gain begins from zero for $V_{in} \approx V_{TH}$ (that is, $g_m \approx 0$) and monotonically increases).

As the drain current increases and g_m increases, A_v approaches $\dfrac{g_m}{g_m + g_{mb}} = \dfrac{1}{1+\eta}$.

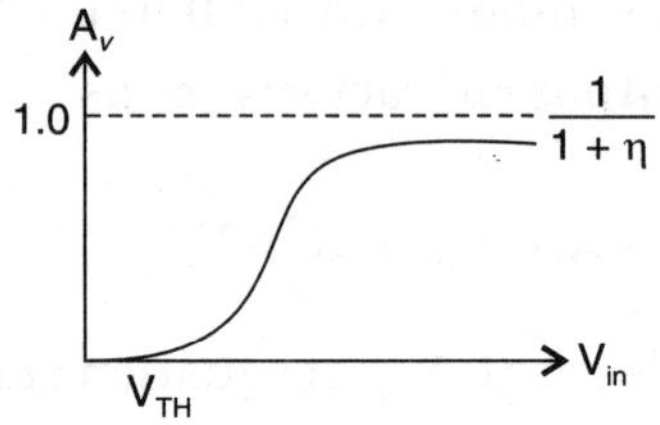

Voltage gain of source follower V_s input voltage

Since $\dfrac{1}{\eta}$ itself slowly decreases with V_{out}, A_v would eventually become equal to unity, but for typical allowable source-bulk voltages, η remains greater than roughly 0.2.

If $R_s = \alpha$, the voltage gain of a source follower is not equal to one.

If $V_{in} < V_{DD}$ then M_1 does not enter the triode region.

In the source-follower, the drain current of M_1 heavily depends on the input d.c. level. To eliminate this issue, the resistor can be replaced by a current source as shown in figure given below :

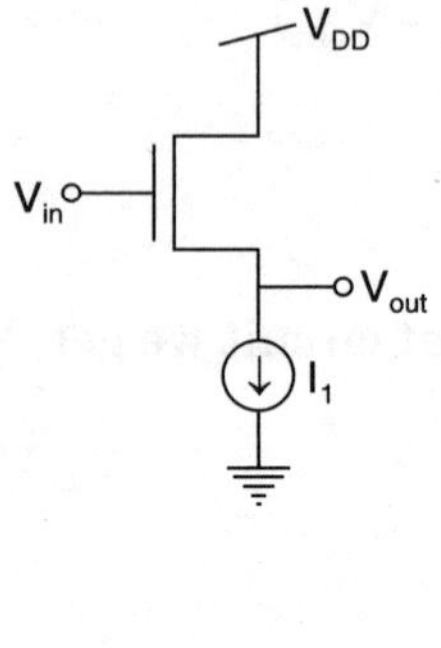

The current source itself is implemented as an nMOS transistor operating in the saturation region, shown in figure given below :

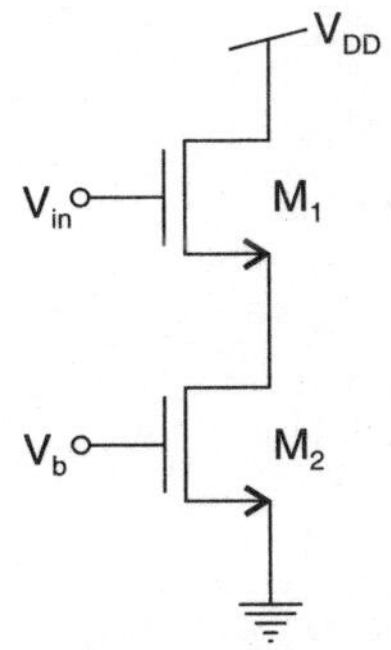

By replacing the R_s by the current source, then the small signal equivalent circuit is,

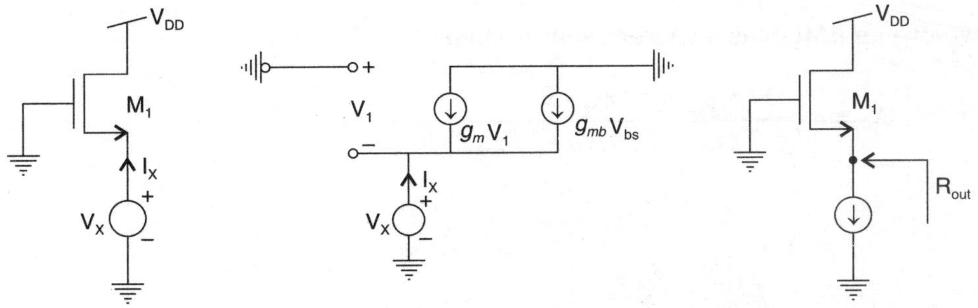

(*a*) (*b*) (*c*)

Calculation of the o/p impedance of a source follower.

From the fig :

$$I_X - g_m V_X - g_{mb} V_X = 0.$$

$$\therefore \qquad R_{out} = \frac{V_X}{I_X} = \frac{1}{g_m + g_{mb}}.$$

From the above equation we can say that, the body effect decreases the o/p resistance of source followers.

With no body effect, only the gate-source voltage of M_1 would increase byAv. With body-effect, on the other hand, the threshold voltage of the device decreases.

Thus in $(V_{as} - V_{Th})^2$, the V_{as} increases but V_{Th} decreases, resulting in a greater change in the drain current and hence a lower o/p impedance.

The magnitude of current source in fig. (*b*), $g_{mb} V_{bs}$ is linearly proportional to the voltage across it.

Such behaviour is that of a simple resistor equal to $\frac{1}{g_{mb}}$, yielding the small signal model which is parallel with the output to lowering the output resistance.

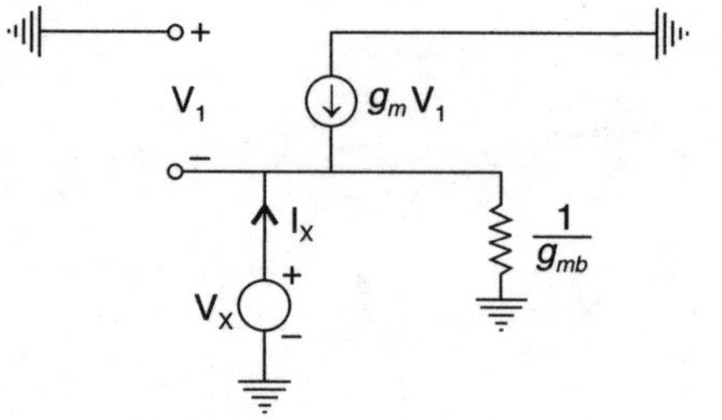

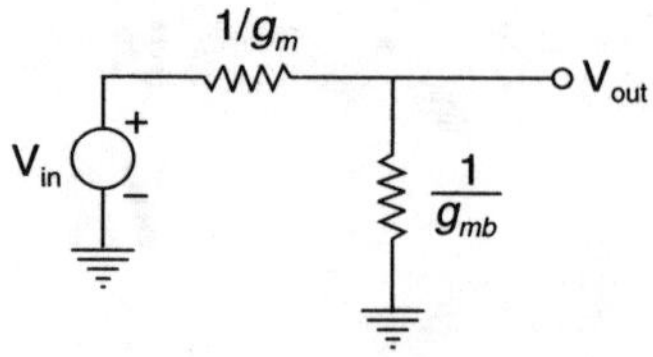

$$\text{So,} \quad R_{out} = \frac{1}{g_m} \parallel \frac{1}{g_{mb}} = \frac{1}{g_m + g_{mb}}.$$

$$\text{So,} \quad A_v = \frac{1/g_{mb}}{1/g_m + 1/g_{mb}} = \frac{g_m}{g_m + g_{mb}}.$$

If we use an nMOS as a current source then,

$$A_v = \frac{1/g_{mb} \parallel r_{O_1} \parallel r_{O_2} \parallel R_L}{1/g_{mb} \parallel r_{O_1} \parallel r_{O_2} \parallel R_L + 1/g_m}.$$

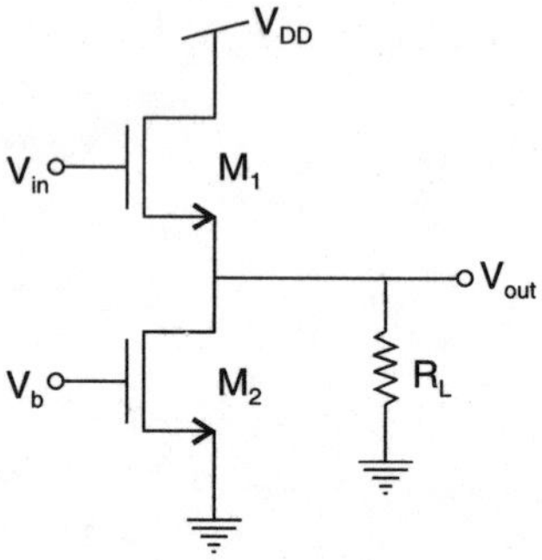

⇒

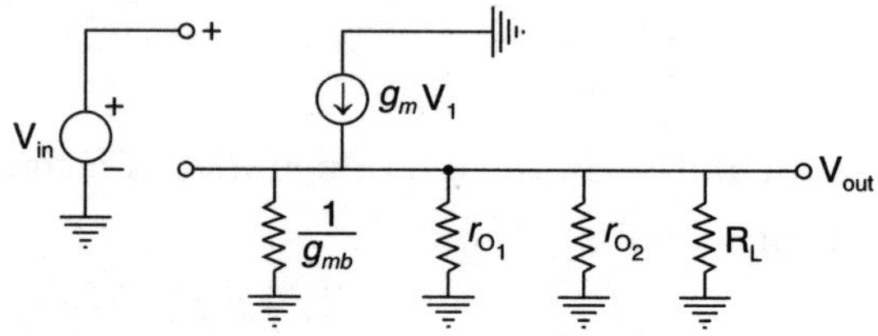

Source follower driving load resistance. Small-signal equivalent circuit.

Advantages : (1) exhibit a high input impedance.

(2) exhibit a moderate output impedance.

Disadvantages : (1) non linearity.

(2) voltage headroom limitation.

Chapter 10 SWITCHED CAPACITOR CIRCUITS AND FILTER DESIGN

10.1 WHAT IS SWITCHED CAPACITOR?

Current through R, $I_R = \dfrac{V_A - V_B}{R}$ ...(1)

The role of the resistor is to take a certain amount of charge from node A at every second and move it to the node B.

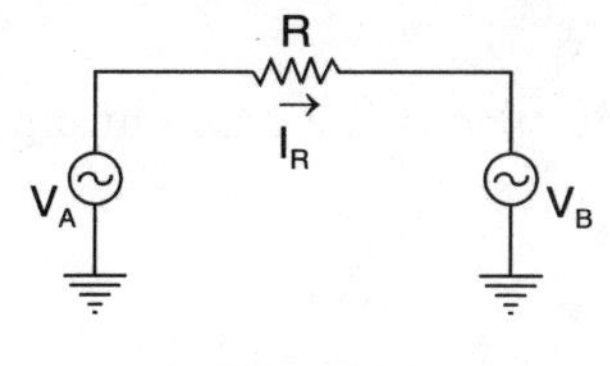

Fig. 10.1

This action can be performed by the capacitor also.

The capacitor c_S is alternately connected to nodes A and B at a rate of f_{ck}/second (clock rate). The charge moves/second is,

$$q = cV.$$

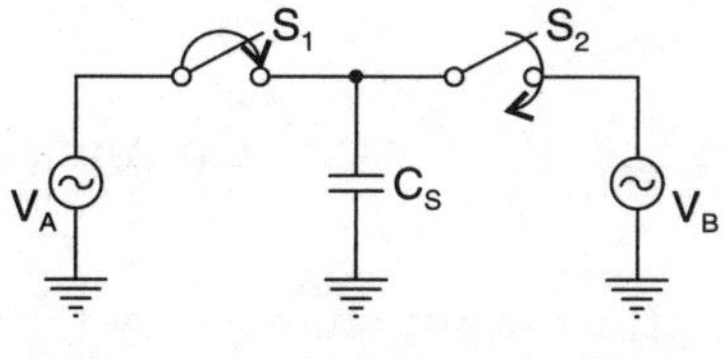

Fig. 10.2. Switched capacitor

or, $$i = \frac{q}{E} = f_{ck} \cdot q.$$

$\therefore$ Average current, $I_{av} = f_{ck}\, C_S\, (V_A - V_B)$ ...(2)

Comparing equations (1) and (2), we note that R can be replaced by a capacitor C_s such that,

$$R = \frac{1}{f_{ck} \cdot C_S} \qquad ...(3)$$

This is the foundation of many switched-capacitor circuits.

10.2 SWITCHED CAPACITOR INTEGRATOR

A : Continuous Time Integrator : Very important component of filters and over sampled A/D converters, is the continuous time integrator. Figure 10.3 shows the circuit.

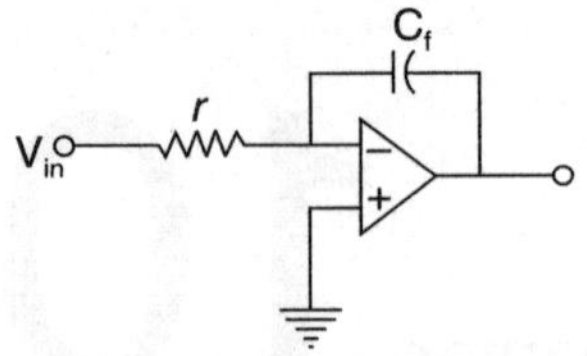

Fig. 10.3

The output is,

$$V_{out} = -\frac{1}{R\,c_f}\int V_{in}\,dt \qquad ...(4)$$

(if the output Amp. gain is very large).

For sampled data system, we need a discrete-time counterpart of this circuit.

B : Discrete Time Integrator :

Here, we replace *r* by a switched capacitor.

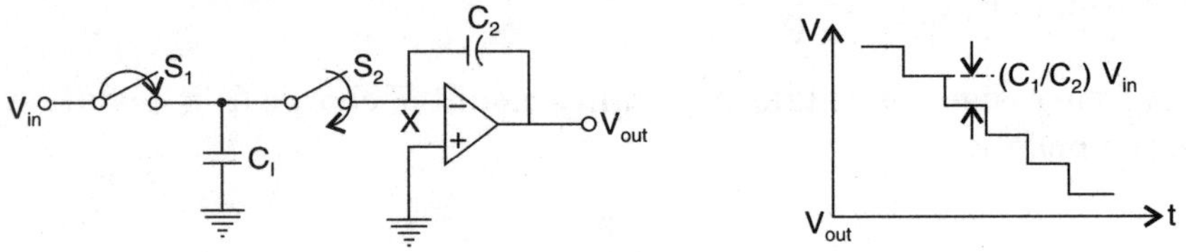

Fig. 10.4. Response to a constant input voltage

$$Z_1 = \frac{1}{j\omega c_1}\,.\ Z_2 = \frac{1}{j\omega c_2}$$

$$V_{out} = -\frac{Z_2}{Z_1}V_{in} = -\frac{1}{j\omega c_2}\cdot\frac{j\omega c_1}{1}V_{in} = -\left|\frac{c_1}{c_2}\right|V_{in}$$

Operation : In every clk pulse,

(1) When S_1 is 'ON' ;

c_1 absorbs a charge $c_1 V_{in}$.

(2) When S_2 is 'ON'

Deposite this charge on c_2. (∵ Op-Amp. input does not draw any current).

Therefore, if V_{in} is constant, the output changes by $\left(\frac{c_1}{c_2}\right)$ V_{in}/clock cycle.

The output waveform is a staircase function. If we can approximate this staircase by a ramp, we get an integrator [c_2 must be large capacitor.]

The output at the end of the *k*th clock cycle is,

$$V_{out}(k\,T_{ck}) = V_{out}[(k-1)\,T_{ck}] - V_{in}[(k-1)\,T_{ck}]\,\frac{c_1}{c_2} \qquad ...(5)$$

Drawbacks of this Discrete-Time Integrator:

(*i*) This input dependent charge injection of S_1 introduces non-linearity in the charge stored in c_1 and hence non-linearity in the output.

(*ii*) The non-linear capacitance at node P resulting from the source/drain junction of S_1 and S_2 leads to a non-linear charge-to-voltage conversion when c_1 is switched to X. This is explained at Fig. 10.5.

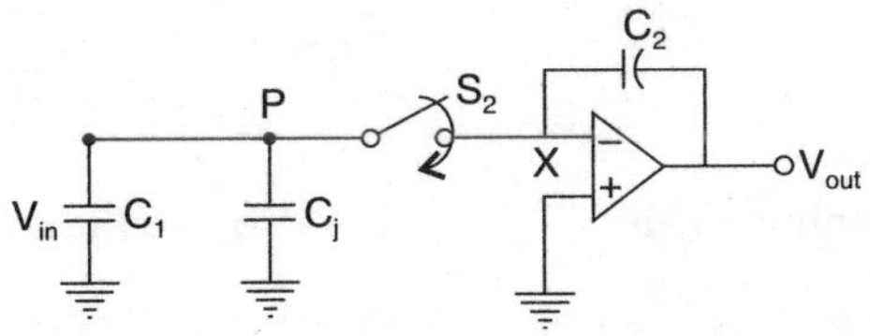

Fig. 10.5. Effect of junction capacitance non-linearity in *sc* Integrator

c_j is junction capacitance. The charge stored in the capacitance c_j is,

$$q_{c_j} = \int_0^{V_{ino}} c_j \,.\, dV \quad ...(6)$$

c_j is a function of V_j, qc_j has non-linear dependence on V_{ino}. This creates a non-linear dependence of the output voltage after the charge is transferred to the integration capacitor.

Solution of this drawback:

Isolate c_1 in the sampling mode. This is shown given below.

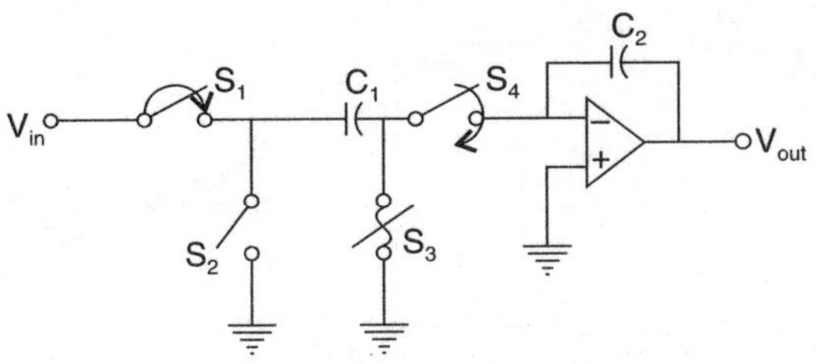

Fig. 10.6(*a*). Parasitic-intensive Integrator

(*i*) **The sampling mode :** S_1 and S_3 are 'ON', S_2 and S_4 are 'OFF'. Voltage across c_1 tracks V_{in}.

Op-Amp. and C_2 holds the precious value.

(*ii*) **The integrating mode :** S_3 turns off first, c_1 gets isolated Fig. 10.6(*b*).

S_1 turns off next. Then V_{in} is disconnected.

S_2 turns 'ON' (the capacitor c_1 gets switched).

S_4 turns 'ON' (the voltage across c_1 is imputed to the op-Amp.).

c_2 gets charged.

Thus the charge on c_1 gets transferred to c_2 through the virtual ground mode.

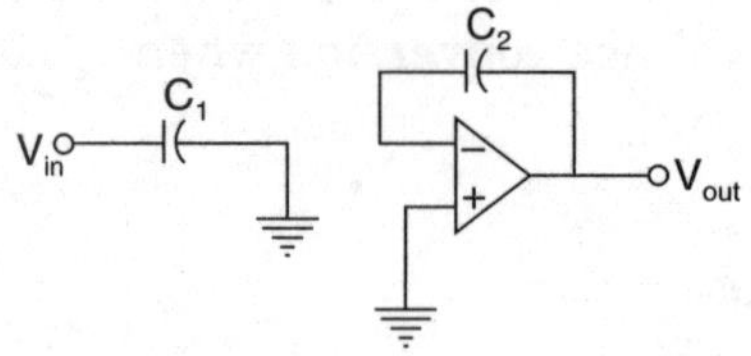

Fig. 10.6(b). Sampling mode

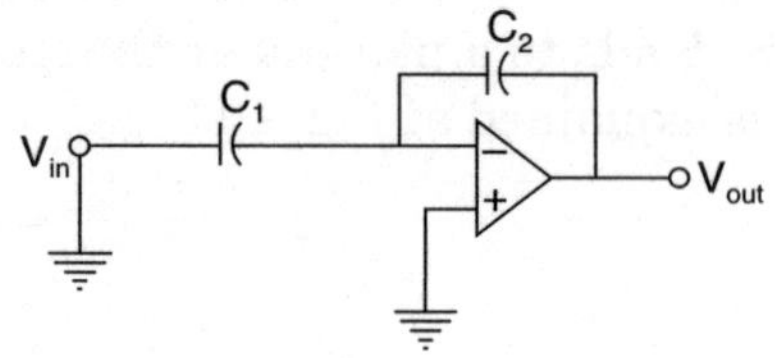

Fig. 10.6(c). Integration mode

Solution of the Problem :

(*i*) Since S_3 turns 'off' first, it introduces only a constant offset. This can be suppressed by differential operation.

(*ii*) Since the node X is a virtual ground, the charge injected or absorbed by S_4 is constant and independent of v_{in}.

10.3 INDEPENDENT VOLTAGE SOURCES

(Z-domain equivalent circuits for a independent voltage source).

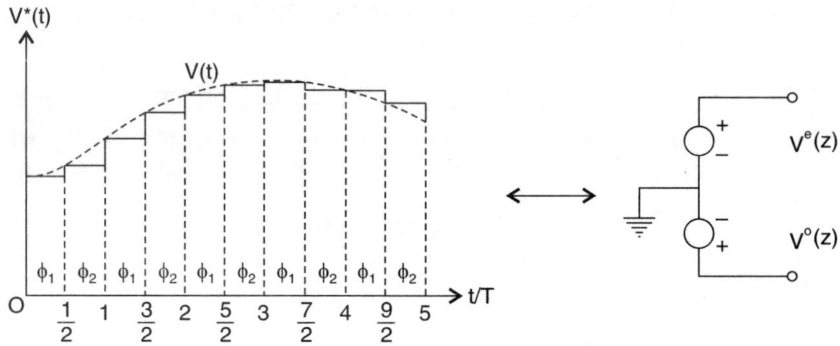

Fig. 10.7. Phase dependent voltage source

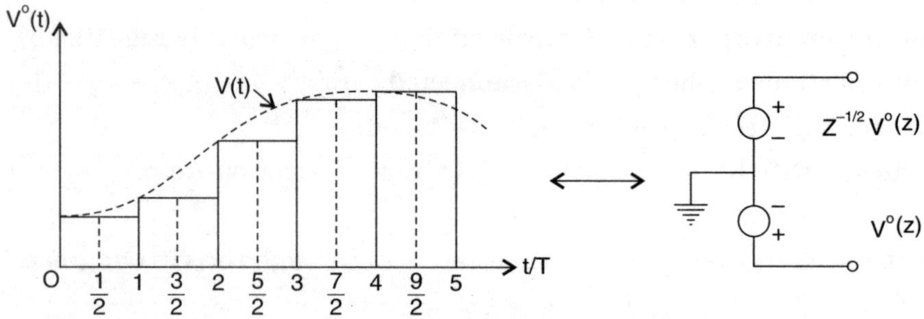

Fig. 10.8. A phase independent voltage source for the odd phase

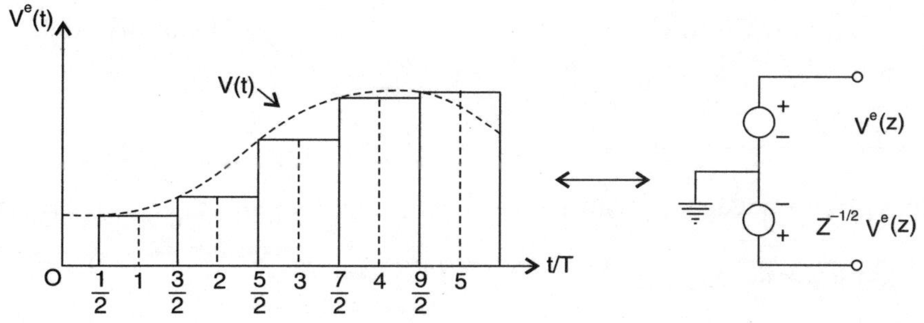

Fig. 10.9. A phase-independent voltage source for the even phase

The values of there independent source depend on the phase of the clock. z-domain, phase-independent models of this voltage source for the odd and even phases are shown in Fig. 10.8 and Fig. 10.9 respectively, along with their waveforms.

Note that the phase-independent voltage sources change value every clock period, T, so that,

$V^e(z) = z^{-1/2} V^0(z)$ and $V^0(z) = z^{-1/2} V^e(z)$ for Fig. 10.8 and 10.9, respectively.

10.4 SWITCHED CAPACITOR TWO-PART CIRCUITS

There are, the z-domain equivalent circuits for four different types of switched capacitor two part circuits are exist. The z-domain models will contain three types of admittances.

We know that, the voltage across the admittance is the stimulus and the current through the admittance is the response.

The first type of admittance is expressed as,

$$I(z) = Yz^{\circ}v(z) = Yv(z).$$

which is interpreted as a current $I(z)$ of value $Yv(z)$ occurs with no delay when a voltage $V(Z)$ is applied.

The second type of admittance is given as,

$$I(z) = Y \,.\, z^{-1/2} v(z).$$

which is interpreted as a current $I(z)$ of value $Yv(z)$ occuring a half period after the voltage $v(z)$ is applied.

The third type of admittance is given as,

$$I(z) = (1 - z^{-1})\ Y\, v(z).$$

which is interpreted as a current that first appears as a value of $Y\,v(z)$ and at a period later this current is zero.

The admittance factor, Y, is equal to the value of the capacitor, c, that is being switched in all three cases.

Switched capacitor Two-Part Circuits. | Simplified Two-part z-Domain Model.

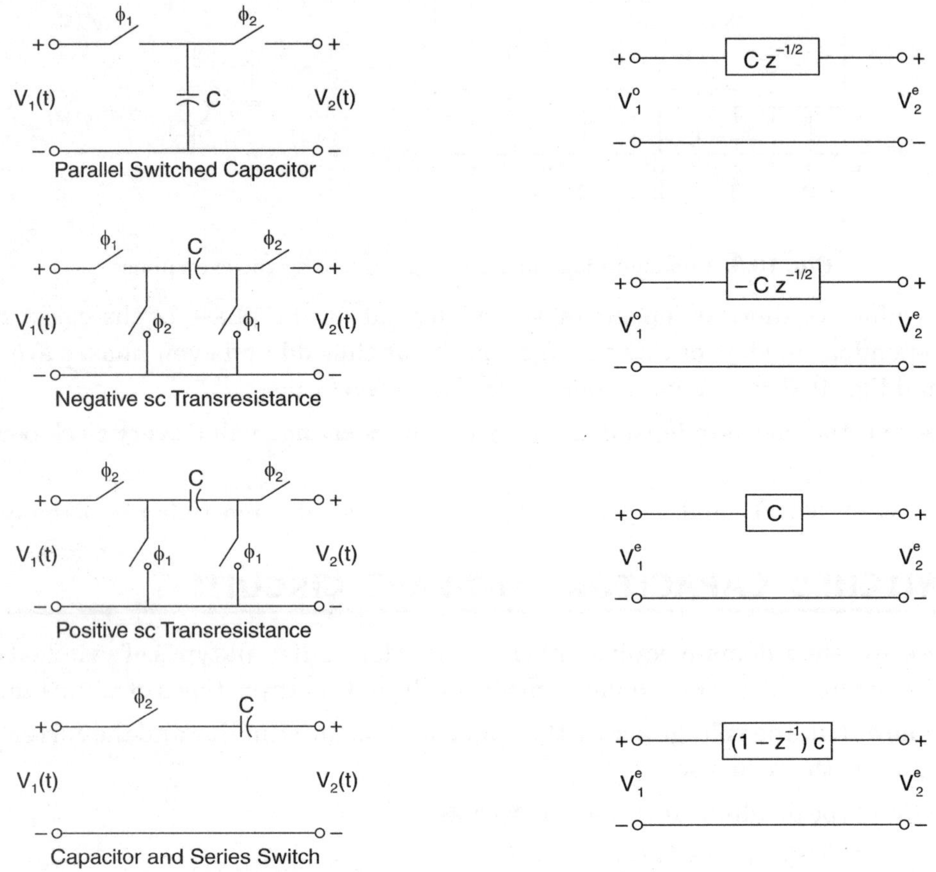

10.5 FIRST ORDER LOW PASS FILTER

A general 1st order transfer function in the s-domain is given as,

$$H(s) = \frac{sa_1 \pm a_0}{s + b_0} \quad ...(1)$$

1st order transfer function has one pole and one zero.

If $a_1 = 0$ then the TF is low pass.

If $a_0 = 0$ then the TF is high pass.

If neither a_0, nor a_1 is zero, the TF is a general 1st order filter.

Note that the zero can be in the RHP or LHP of the complex frequency domain.

So, in z-domain,

$$H(z) = \frac{zA_1 \pm A_0}{z - B_0} = \frac{A_1 \pm A_0 z^{-1}}{1 - B_0 z^{-1}} \quad ...(2)$$

LPF :

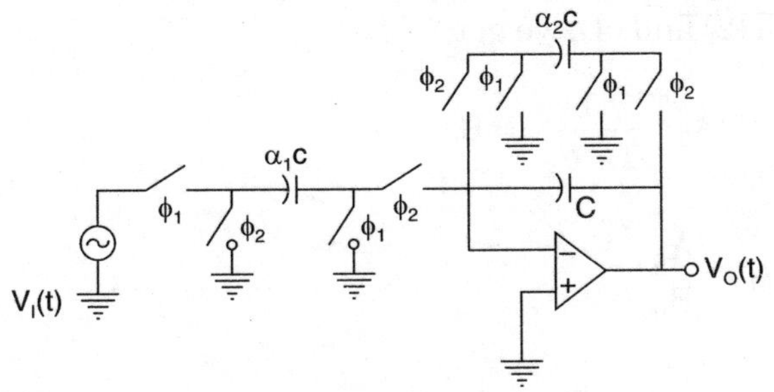

Inverting 1st order LPF.

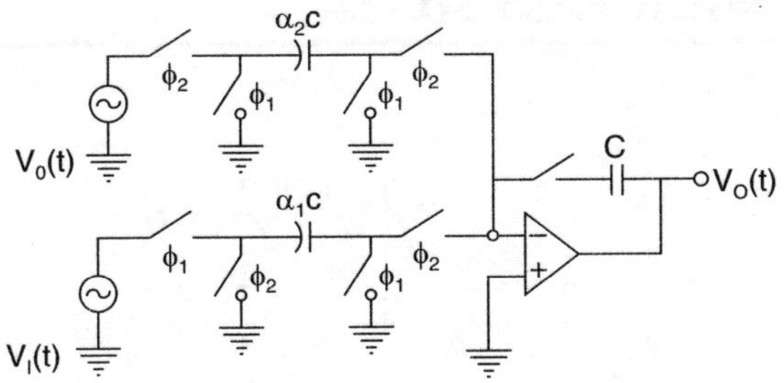

Modified 1st order LPF equivalent circuit.

$V_0^{\,0} = z^{-1/2}\, V_0^{\,e}$.

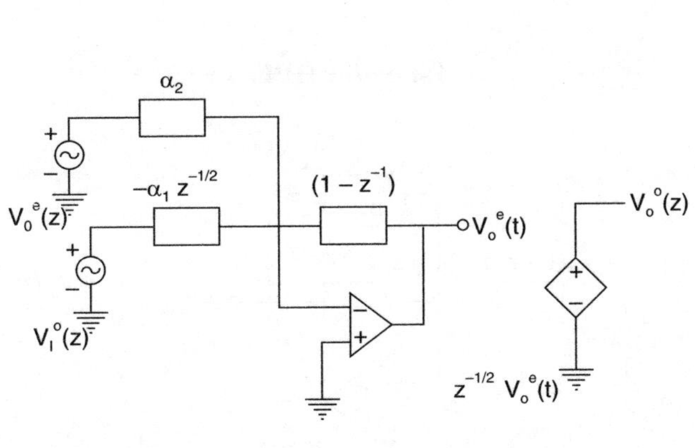

Z domain model.

Summing currents flowing toward the inverting op-amp terminal gives,

$$\alpha_2 c_1 V_0^e(z) - \alpha_1 c_1 z^{-1/2} V_i^0(z) + c_1(1 - z^{-1}) V_0^e(z) = 0 \quad \text{...(3)}$$

$$\therefore \quad \frac{V_0^0(z)}{V_i^0(z)} = \frac{z^{-1/2} V_0^e(z)}{V_i^0(z)} = \frac{\alpha_1 z^{-1}}{1 + \alpha_2 - z^{-1}} = \frac{\alpha_1 z^{-1}/1 + \alpha_2}{1 - \dfrac{z^{-1}}{1 + \alpha_2}} \quad \text{...(4)}$$

Equating equation (2) and (4), we get,

$$\frac{\alpha_1}{1 + \alpha_2} = A_0 \cdot \frac{1}{1 + \alpha_2} = B_0$$

$$\alpha_1 = \frac{A_0}{B_0}$$

$$\alpha_2 = \frac{1 - B_0}{B_0}. \quad \text{...(5)}$$

10.6 FIRST ORDER HIGH PASS FILTER

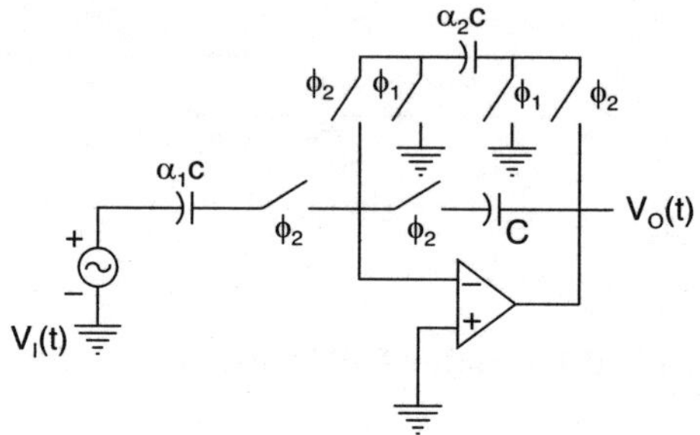

1st order HPF.

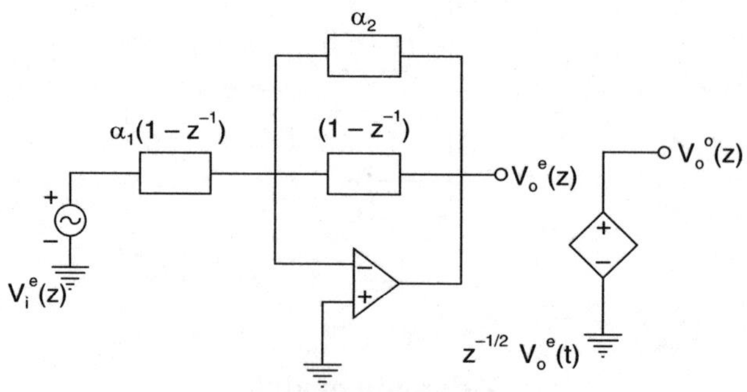

Z-domain Model.

Summing currents at the inverting input node of the OP-Amp gives,

$$\alpha_1 (1 - z^{-1}) V_i^e (z) + \alpha_2 V_0^e (z) + (1 - z^{-1}) V_0^e (z) = 0 \quad ...(6)$$

$$H(Z) = \frac{V_0^e (z)}{V_i^e (z)} = \frac{-\alpha_1 (1 - z^{-1})}{\alpha_2 + (1 - z^{-1})} = \frac{\dfrac{\alpha_1}{\alpha_1 + 1} (1 - z^{-1})}{1 - \dfrac{z^{-1}}{1 + \alpha_2}} \quad ...(7)$$

Equating equations (2) and (7), we get

$$\alpha_1 = -\frac{A_1}{B_0} = \frac{-A_0}{B_0} \quad \text{and} \quad \alpha_2 = \frac{1 - B_0}{B_0}. \quad ...(8)$$

APPENDIX

SPECIAL ARITHMETIC VLSI CIRCUITS

A CARRY LOOKAHEAD ADDER

A **carry lookahead adder** is a type of adder used in digital logic. It can be contrasted with the simpler, but usually slower, *ripple carry adder.*

A ripple-carry adder works in the same way as pencil-and-paper methods of addition. Starting at the rightmost (least significant) digit position, the two corresponding digits are added and a result obtained. It is also possible that there may be a carry out of this digit position (for example, in pencil-and-paper methods, "9 + 5 = 4, carry 1"). Accordingly all digit positions other than the rightmost need to take into account the possibility of having to add an extra 1, from a carry that has come in from the next position to the right.

This means that no digit position can have an absolutely final value until it has been established whether or not a carry is coming in from the right. Moreover, unless the sum without a carry is 9 (in pencil-the-paper methods) or 1 (in binary arithmetic), it is not even possible to tell whether or not a given digit position is going to pass on a carry to the position on its left. At worst, when a whole sequence of sums comes to ...99999999... (in decimal) or ...11111111... (in binary), nothing can be deduced at all until the value of the carry coming in from the right is known, and that carry is then propagated to the left, one step at a time, as each digit position evaluated "9 + 1 = 0, carry 1" or "1 + 1 = 0, carry 1". It is the "rippling" of the carry from right to left that gives a ripple-carry adder its name, and its slowness. When adding 32-bit integers, for instance, allowance has to be made for the possibility that a carry could have to ripple through everyone of the 32 one-bit adders.

Carry lookahead depends on two things :

1. Calculating, for each digit position, whether that position is going to propagate a carry if one comes in from the right.
2. Combining these calculated values so as to be able to deduce quickly whether, for each group of digits, that group is going to propagate a carry that comes in from the right.

Supposing that groups of 4 digits are chosen. Then the sequence of events goes something like this :

1. All 1-bit address calculate their results. Simultaneously, the lookahead units perform their calculations.
2. Suppose that a carry arises in a particular group. Within at most 3 gate delays, that carry will emerge at the left-hand end of the group and start propagating through the group to its left.

3. If that carry is going to propagate all the way through the next group, the lookahead unit will already have deduced this. Accordingly, *before the carry emerges from the next group* the lookahead unit is immediately (within 1 gate delay) able to tell the *next* group to the left that it is going to receive a carry-and, at the same time, to tell the next lookahead unit to the left that a carry is on its way.

The net effect is that the carries start by propagating slowly through each 4-bit group, just as in a ripple-carry system, but then move 4 times as fast, leaping from one lookahead carry unit to the next. Finally, within each group that receives a carry, the carry propagates slowly within the digits in that group.

The more bits in a group, the more complex the lookahead carry logic becomes, and the more time is spent on the "slow roads" in each group rather than on the "fast road" between the groups (provided by the lookahead carry logic). On the other hand, the fewer bits there are in a group, the more groups have to be traversed to get from one end of a number to the other, and the less acceleration is obtained as a result.

Deciding the group size to be governed by lookahead carry logic requires a detailed analysis of gate and propagation delays for the particular technology being used.

It is possible to have more than one level of lookahead carry logic, and this is in fact usually done. Each lookahead carry unit already produces a signal saying "if a carry comes in from the right, I will propagate it to the left", and those signals can be combined so that each group of (let us say) four lookahead carry units becomes part of a "supergroup" governing a total of 16 bits of the numbers being added. The "supergroup" lookahead carry logic will be able to say whether a carry entering the supergroup will be propagated all the way through it, and using this information, it is able to propagate carries from right to left 16 times as fast as a naive ripple carry. With this kind of two-level implementation, a carry may first propagate through the "slow road" of individual adders, then, on reaching the left-hand end of its group, propagate through the "fast road" of 4-bit lookahead carry logic, then, on reaching the left-hand end of its supergroup, propagate through the "superfast road" of 16-bit lookahead carry logic.

Again, the group sizes to be chosen depend on the exact details of how fast signals propagate within logic gates and from one logic gate to another.

For very large numbers (hundreds or even thousands of bits) lookahead carry logic does not become any more complex, because more layers of supergroups and supersupergroups can be added as necessary. The increase in the number of gates is also moderate: if all the group sizes are 4, one would end up with one third as many lookahead carry units as there are adders. However, the "slow roads" on the way to the faster levels begin to impose a drag on the whole system (for instance, a 256-bit adder could have up to 24 gate delays in its carry processing), and the mere physical transmission of signals from one end of a long number to the other begins to be a problem. At these sizes carry-save adders are perferable, since they spend no time on carry propagation at all.

Carry Lookahead Method

Carry lookahead logic uses the concepts of *generating* and *propagating* carries. Although in the context of a carry lookahead adder, it is most natural to think of generating and propagating in the context of binary addition, the concepts can be used more generally than this. In the descriptions below, the word *digit* can be replaced by *bit* when referring to binary addition.

The addition of two 1-digit inputs A and B is said to *generate* if the addition will always carry, regardless of whether there is an input carry (equivalently, regardless of whether any less significant digits in the sum carry). For example, in the decimal addition 52 + 67, the addition of the tens digits 5 and 6 *generates* because the result carries to the hundreds digit regardless of whether the ones digit carries (in the example, the ones digit does not carry (2 + 7 = 9)).

In the case of binary addition, A + B generates if and only if both A and B are 1. If we write G(A, B) to represent the binary predicate that is true if and only if A + B generates, we have :

$$G(A, B) = A . B$$

The addition of two 1-digit inputs A and B is said to *propagate* if the addition will carry whenever there is an input carry (equivalently, when the next less significant digit in the sum carries). For example, in the decimal addition 37 + 62, the addition of the tens digits 3 and 6 *propagate* because the result would carry to the hundreds digit *if* the ones were to carry (which in this example, it does not). Note that propagate and generate are defined with respect to a single digit of addition and do not depend on any other digits in the sum.

In the case of binary addition, A + B propagates if and only if at least one of A or B is 1. If we write P(A, B) to represent the binary predicate that is true if and only if A + B propagates, we have :

$$P(A, B) = A + B$$

Sometimes a slightly different definition of *propagate* is used. By this definition A + B is said to propagate if the addition will carry whenever there is an input carry, but will not carry if there is no input carry. It turns out that the way in which generate and propagate bits are used by the carry lookahead logic, it doesn't matter which definition is used. In the case of binary addition, this definition is expressed by :

$$P'(A, B) = A \oplus B$$

For binary arithmetic, *or* is faster than *xor* and takes fewer transistors to implement. However, for a multiple-level carry lookahead adder, it is simpler to use P′(A, B).

Given these concepts of generate and propagate, when will a digit of addition carry ? It will carry precisely when either the addition generates *or* the next less significant bit carries and the addition propagates. Written in boolean algebra, with C_i the carry bit of digit i, and P_i and G_i the propagate and generate bits of digit i respectively,

$$C_{i+1} = G_i + (P_i . C_i)$$

Implementation Details

For each bit in a binary sequence to be added, the Carry Look Ahead Logic will determine whether that bit pair will generate a carry or propagate a carry. This allows the circuit to "pre-process" the two numbers being added to determine the carry ahead of time. Then, when the actual addition is performed, there is no delay from waiting for the ripple carry effect (or time it takes for the carry from the first Full Adder to be passed down to the last Full Adder). Below is a simple 4-bit generalized Carry Look Ahead circuit that combines with the 4-bit Ripple Carry Adder we used above with some slight adjustments :

For the example provided, the logic for the generate (g) and propagate (p) values are given below. Note that the numeric value determines the signal from the circuit above, starting from 0 on the far left to 3 on the far right:

$$C_1 = G_0 + P_0 . C_0$$
$$C_2 = G_1 + P_1 . C_1$$
$$C_3 = G_2 + P_2 . C_2$$
$$C_4 = G_3 + P_3 . C_3$$

Substituting C_1 into C_2, then C_2 into C_3, then C_3 into C_4 yields the expanded equations :

$$C_1 = G_0 + P_0 . C_0$$
$$C_2 = G_1 + G_0 . P_1 + C_0 . P_0 . P_1$$
$$C_3 = G_2 + G_1 . P_2 + G_0 . P_1 . P_2 + C_0 . P_0 . P_1 . P_2$$
$$C_4 = G_3 + G_2 . P_3 + G_1 . P_2 . P_3 + G_0 . P_1 . P_2 . P_3 + C_0 . P_0 . P_1 . P_2 . P_3$$

To determine whether a bit pair will generate a carry, the following logic works :

$$G_i = A_i . B_i$$

To determine whether a bit pair will propagate a carry, either of the following logic statements work :

$$P_i = A_i \oplus B_i$$
$$P_i = A_i + B_i$$

The reason why this works is based on evaluation of $C_1 = G_0 + P_0 . C_0$. The only difference in the truth tables between (A ⊕ B) and (A + B) is when both A and B are 1. However, if both A and B are 1, then the G_0 term is 1 (since its equation is A . B), and the P_0. C_0 term becomes irrelevant. The XOR is used normally within a basic full adder circuit; the OR is an alternate option (for a carry lookahead only) which is far simpler in transistor-count terms.

The Carry Look Ahead 4-bit adder can also be used in a higher-level circuit by having each CLA Logic circuit produce a propagate and generate signal to a higher-level CLA logic circuit. The group propagate (PG) and group generate (GG) for a 4-bit CLA are :

$$PG = P_0 . P_1 . P_2 . P_3$$
$$GG = G_3 + G_2 . P_3 + G_1 . P_3 . P_2 + G_0 . P_3 . P_2 . P_1$$

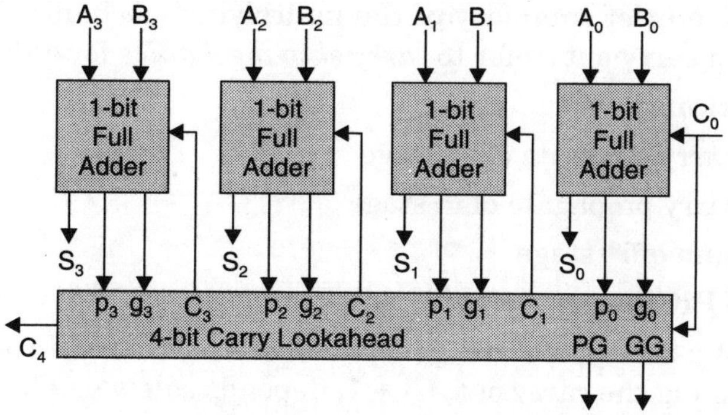

Fig. 1. 4-bit adder with LCU

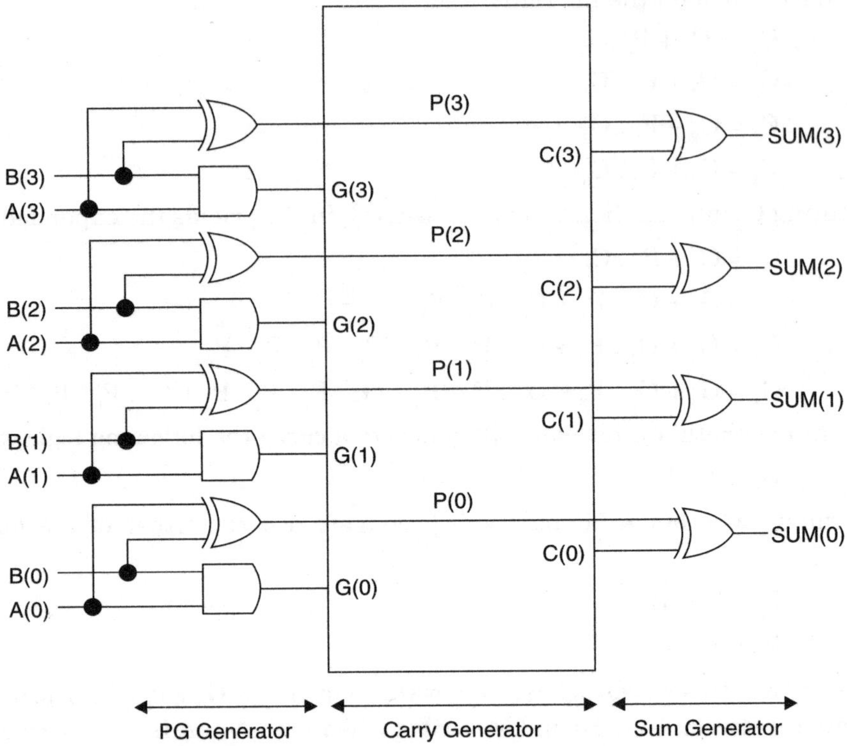

Fig. 2. 4-bit Carry Lookahead Adder

MANCHESTER CARRY ADDER

The propagation time, when calculating the sum of two binary strings **A** and **B** using any generic parallel adder, can be speed up significantly if we utilize a **Manchester cell** in the design of that particular adder. Alternatively one may choose to perform the addition using any of the two flavours of **Manchester adders** described later in this section.

Generation and Propagation

Here we provide a brief summary of the underlying mechanics behind the decision to propagate or generate a carry out (refer to *carry skip mechanics* for a through explanation).

Boolean Equations :

1. **Gi = Ai Bi**—carry generate of ith stage
2. **Pi = Ai Bi**—carry propagate of ith stage
3. **Si = Pi Ci**—sum of ith stage
4. **Ci + 1 = Gi + PiCi**—carry out of ith stage

The condition for a carry generate (generation of a new carry) to occur at any stage of the addition of **Ai = Bi** making the carry out, **Ci + 1,** depends solely on **Gi** (*i.e.,* **Ci + 1 = Gi**—from equation 4). A carry propagate, on the other hand, has the requirement that ***Ai Bi***, hence producing ***Ci + 1 = Ci*** from equation 4. Figure (*a*), (*b*) and (*c*) describe three possible transistor level implementations for a single carry propagate cell as known as a **Manchester cell** (all of these versions implement equation 4 listed above with as little transistors as possible without compromising speed and performance).

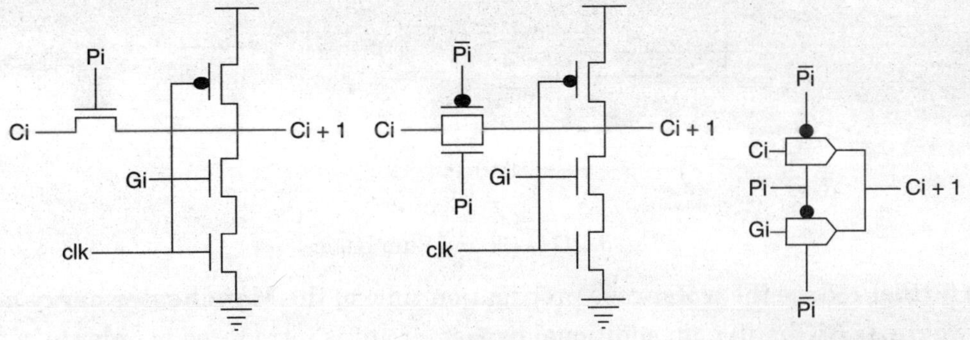

(*a*) Dynamic Stage (*b*) Static Stage (*c*) MUX Stage

Fig. 3

A **Manchester carry adder** consists of cascaded stages of Manchester propagation cells, shown above. The optimum amount of cascaded stages may be calculated for a technology by simulation. For a 16 bit adder example a 4-bit adder made up of four static stage cells, shown in figure 4, is chosen in order to reduce the number of series-propagate transistors, which greatly improves speed. In the case of a four-bit adder, the maximum number of transistors that are in series with the gate, when all propagate signals and Ci is true, is only five.

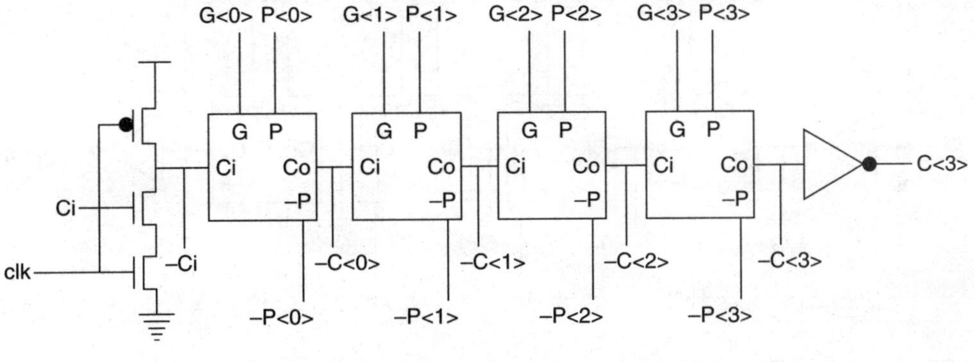

Fig. 4. 4-bit Manchester carry section

In addition to the cascaded **Manchester** propagation cells the adder requires carry propagation and carry generation logic, also called a PG generator shown in figure (5).

Finally to complete the design four XNOR blocks each of which produces the SUM at each particular stage is required.

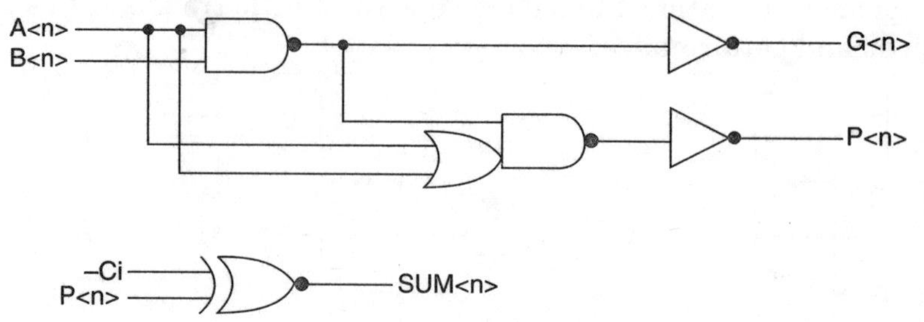

Fig. 5. PG Logic and Sum Logic

To further reduce the worst-case propagation time of the **Manchester carry adder** in the case where ***Ai Bi***, for all *i*, an additional bypass circuit is introduced in order to bypass the four stages. The circuit is illustrated in figure 6.

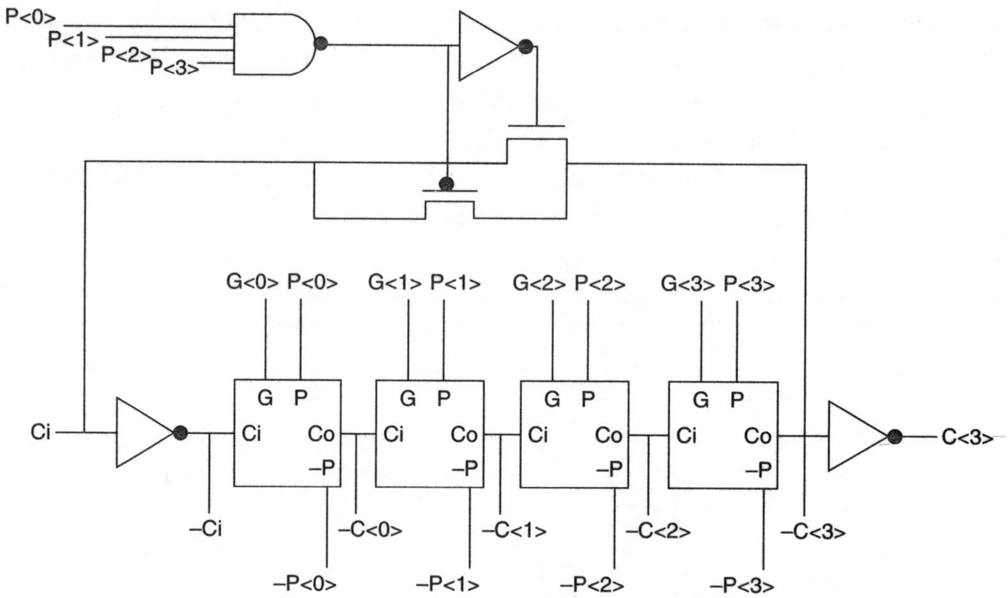

Fig. 6. Manchester Carry adder with Carry bypass

Other **Manchester adders'** implementations are possible. One such adder is based on MUXes called a conflict free Manchester Adder. Although this version reduces even further the propagation time of the adder, it still embodies the core of a Manchester adder whose ultimate goal is to achieve the reduction of the worst-case time propagation by employing a Manchester cell.

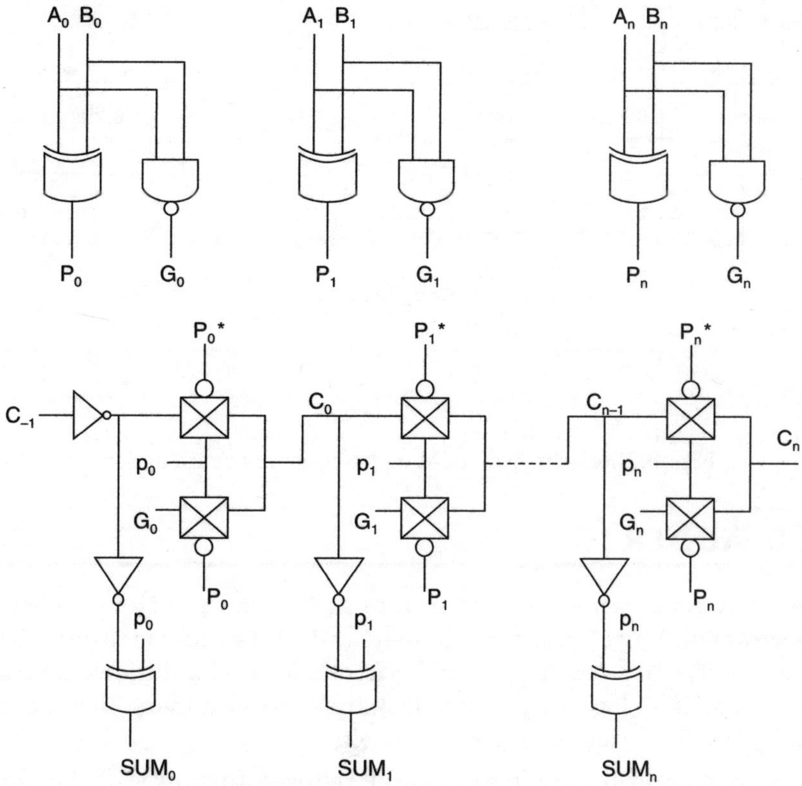

Fig. 7. The Configuration of the Manchester Adder/Subtractor

A Manchester Adder can be constructed by designing a cell and cascading it as shown in the figures below.

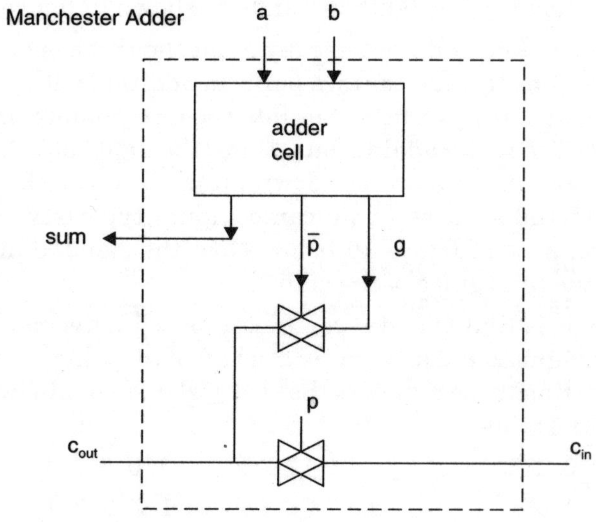

Fig. 8. Manchester Adder Cell

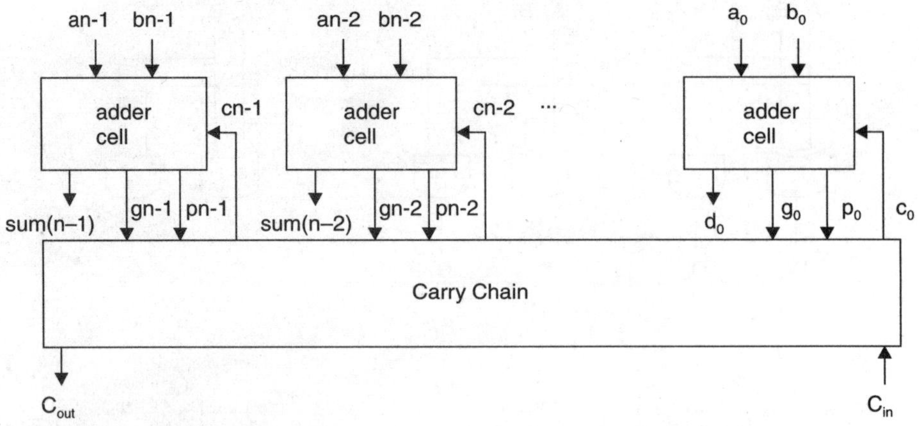

Fig. 9. Block diagram of Manchester carry chain adder

CARRY-SAVE ADDER

There are many cases where it is desired to add more than two numbers together. The straightforward way of adding together m numbers (all n bits wide) is to add the first two, then add that sum to the next, and so on. This requires a total of $m - 1$ additions, for a total gate delay of the order of $m \log n$ or O ($m \log n$), assuming lookahead carry adders. Instead, a tree of adders can be formed, taking only O ($\log m \,.\, \log n$) gate delays.

Using carry-save addition, the delay can be reduced further still. The idea is to take 3 numbers that we want to add together, X + Y + Z, and convert it into 2 numbers C + S such that X + Y + Z = C + S, and do this in O(1) time. The reason why addition cannot be performed in O(1) time is because the carry information must be propagated. In carry save addition, we refrain from directly passing on the carry information until the very last step. We will first illustrate the general concept with a binary two 5-bit operands as shown in figure 10.

To add three numbers by hand, we typically align the three operands, and then proceed column by column in the same fashion that we perform addition with two numbers. The three digits in a row are added, and any overflow goes into the next column. In this case, when there is some non-zero carry, we are really adding four digits (the digits of X, Y and Z, plus the carry). The carry-save approach breaks this process down into two steps. The first is to compute the sum ignoring any carries and separately, we can compute the carry on a column by column basis as shown in the top-row of figure 10 below. Now, the same 'Sum' can be computed by adding the 'S' and 'C' in the final stage of the addition.

The important point is that C and S can be computed independently, and furthermore, each Ci, (and Si) can be computed independently from all of the other C's (and S's). This achieves our original goal of converting three numbers that we wish to add into two numbers that add up to the same sum and in O(1) time.

X:	1 0 0 1 1
Y:	+1 1 0 0 1
Z:	+0 1 0 1 1
S:	0 0 0 0 1

X:	1 0 0 1 1
Y:	+1 1 0 0 1
Z:	+0 1 0 1 1
C:	1 1 0 1 1

X:	1 0 0 1 1
Y:	+ 1 1 0 0 1
Z:	+ 0 1 0 1 1
S:	0 0 0 0 1
C:	1 1 0 1 1
Sum:	1 1 0 1 1 1

Fig. 10. Carry-save addition process of three 5-bit operands

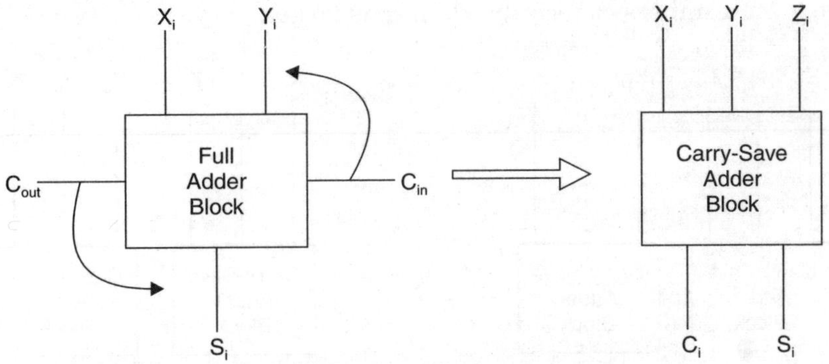

Fig. 11. The carry-save adder block is the same circuit as the full adder

In many cases we need to add several operands together, carry-save adders are ideal for this type of addition. A carry-save adder consists of a ladder of stand alone full adders, and carries out a number of partial additions. The principal idea is that the carry has a higher power of 2 and thus is routed to the next column. Doing additions with carry-save adder saves time and logic.

In this method, for the first 3 numbers a row of full adders are used. Then a row of full adders is added for each additional number. The final results, in the form of two numbers SUM and CARRY, are then summed up with a carry propagate adder or any other adder. An example of 4 numbers addition is shown in figure 12.

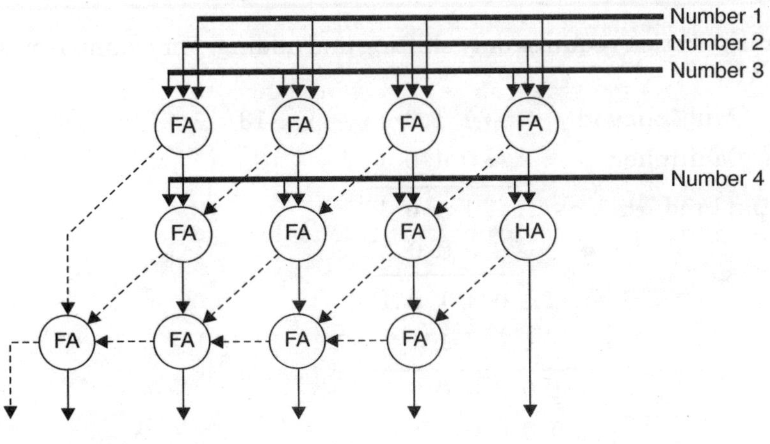

Figure 12. Carry-Save Adder for as four bit number

The carry-save adder structure can be implemented using Full Adders. A full adder, but with some of the signals renamed as shown in Figure 11 can be used to compute 'C' and 'S' in a carry-save adder. A carry-save adder simply is a full adder with the 'C_{in}' input renamed to 'Z_i', the 'C_{out}' output renamed to 'C_i'. Figure 13 shows how sixteen carry save adders can be arranged to add three 16 bit numbers X, Y and Z into two numbers 'C' and 'S'. Note that the CSA block in bit position zero generates 'C_0' which gets added to 'S_1'. Also, the first carry save adders output 'S_0' is the least significant bit of the 'Sum'.

Note that all of the CSA blocks are independent, thus the entire circuit takes only O(1) time. To get the final sum, we still need a Carry Look-Ahead Adder in the last stage, which will cost us O ($\log n$) delay. The gate delay to add three n-bit numbers is thus the nearly the same as adding only two n-bit numbers especially when 'n' is larger.

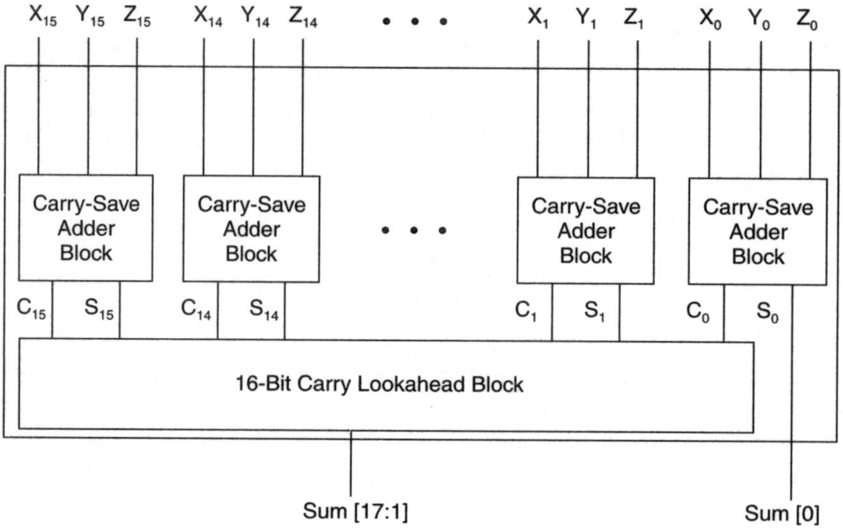

Fig. 13. Three operand 16-Bit Carry-Save Adder Block Diagram

BINARY SERIAL-PARALLEL MULTIPLIER

Binary multiplication requires only shifting and adding. For example, multiplication $(13)_{10}$ by $(05)_{10}$ in binary:

```
          Multiplicand    →1 1 0 1      (13)
          Multiplier      →0 1 0 1      (05)
                          --------
       Partial Product    →1 1 0 1
                           0 0 0 0
                          --------
                           0 1 1 0 1
                         1 1 0 1
                        ----------
                       1 0 0 0 0 0 1
                       0 0 0 0
                      ------------
                     0 1 0 0 0 0 0 1    (65)
```

Note that each partial product is either the multiplicand (1101) shifted over by the appropriate number of places or zero.

Multiplication of two 4-bit numbers requires a 4-bit multiplicand register, a 4-bit multiplier register, a 4-bit full adder, and an 8-bit register for the product. The product register serves as an accumulator to accumulate the sum of the partial products. If the multiplicand were shifted left each time before it was added to the accumulator, as was done in the previous example, an 8-bit adder would be needed. So it is better to shift the contents of the product register to the right each time, as shown in the block diagram of figure 14. This type of multiplier is sometimes referred to as a serial—parallel multiplier, since the multiplier bits are processed serially, but the addition takes place in parallel. As indicated by the arrows on the diagram, 4 bits from the accumulator (ACC) and 4 bits from the multiplicand register are connected to the adder inputs; the 4 sum bits and the carry output from the adder are connected back to the accumulator. When an add signal (Ad) occurs, the adder outputs are transferred to the accumulator by the next clock pulse, thus causing the multiplicand to be added to the accumulator. An extra bit at the left end of the product register temporarily stores any carry that is generated when the multiplicand is added to the accumulator. When a shift signal (Sh) occurs, all 9 bits of ACC are shifted right by the next clock pulse.

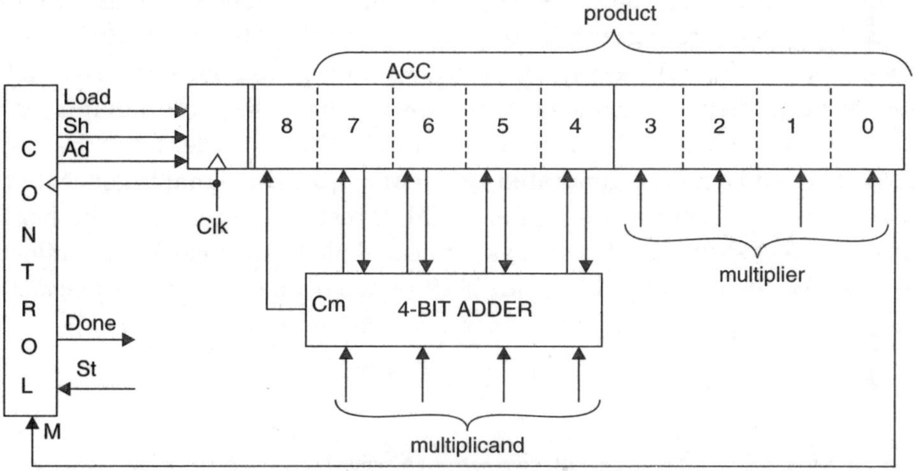

Fig. 14. Block Diagram for Binary Multiplier

Since the lower 4 bits of the product register are initially unused, we will store the multiplier in this location instead of in a separate register. As each multiplier bit is used, it is shifted out the right end of the register to make room for additional product bits. A shift signal (Sh) causes the contents of the product register (including the multiplier) to be shifted right one place when the next clock pulse occurs. The control circuit puts out the proper sequence of add and shift signals after a start signal (St = 1) has been received. If the current multiplier bit (M) is 1, the multiplicand is added to be accumulator followed by a right shift; if the multiplier bit is 0, the addition is skipped, and only the right shift occurs. The multiplication example (13 × 05) is reworked below showing the location of the bits in the registers at each clock time.

Multiplication Steps :

Initial contents of product register	0 0 0 0 0 0 1 0 1 ← M(5)
(add multiplicand since M= 1)	1 1 0 1
after addition	0 1 1 0 1 0 1 0 1
after shift	0 0 1 1 0 1 0 1 0 ← M
(skip addition since M = 0)	0 0 0 0
after addition	0 0 1 1 0 1 0 1 0
after shift	0 0 0 1 1 0 1 0 1 ← M
(add multiplicand since M = 1)	1 1 0 1
after addition	1 0 0 0 0 0 1 0 1
after shift	0 1 0 0 0 0 0 1 0
(skip addition since M = 0)	
after shift (final answer)	0 0 1 0 0 0 0 0 1

The control circuit must be designed to output the proper sequence of add and shift signals. Figure 15 shows a state graph for the control circuit. In figure 15, S0 is the reset state, and the network stays in S0 until a start signal (St = 1) is received. This generates a load signal, which causes the multiplier to be loaded into the lower 4 bits of the Accumulator (ACC) and the upper 5 bits of the accumulator to be cleared. In state S1, the low-order bit of the multiplier (M) is tested. If M = 1, an add signal is generated, and if M = 0, a shift signal is generated. Similarly, in states S3, S5 and S7, the current multiplier bit (M) is tested to determine whether to generate an add or shift signal. A shift signal is always generated at the next clock time following an add signal (states S2, S4, S6, and S8). After four shifts have been generated, the control network goes to S9, and a done signal is generated before returning to S0.

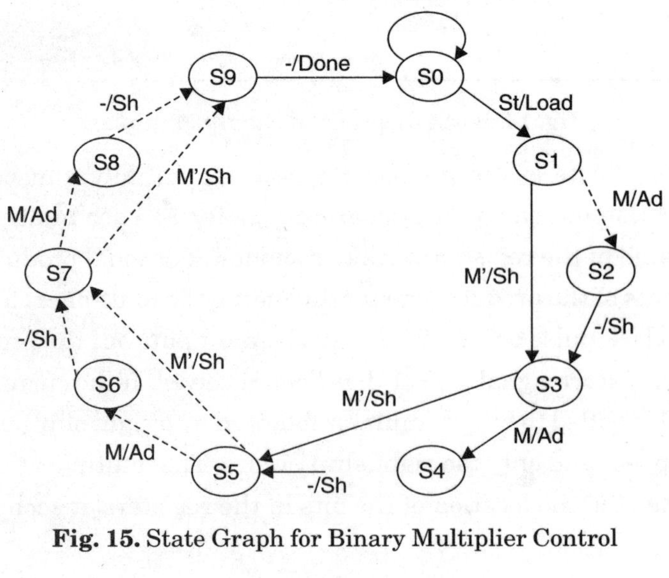

Fig. 15. State Graph for Binary Multiplier Control

As the state graph for the multiplier indicates, the control performs two functions generating add or shift signals as needed and counting the number of shifts. If the number of bits is large, it is convenient to divide the control network into a counter and an add-shift control, as shown in figure 16(*a*). First, we will derive a state graph for the add-shift control that tests St and M and outputs the proper sequence of add and shift signals (figure 16(*b*)). Then we will add a completion signal (K) from the counter that stops the multiplier after the proper number of shifts have been completed. Starting in S0 in figure 16(*b*), when a start signal St = 1 is received, a load signal is generated and the network goes to state S2; if M = 0, a shift signal is generated and the network stays in S1. In S2, a shift signal is generated since a shift always follows an add. The graph of figure 16(*b*) will generate the proper sequence of add and shift signals, but it has no provision for stopping the multiplier.

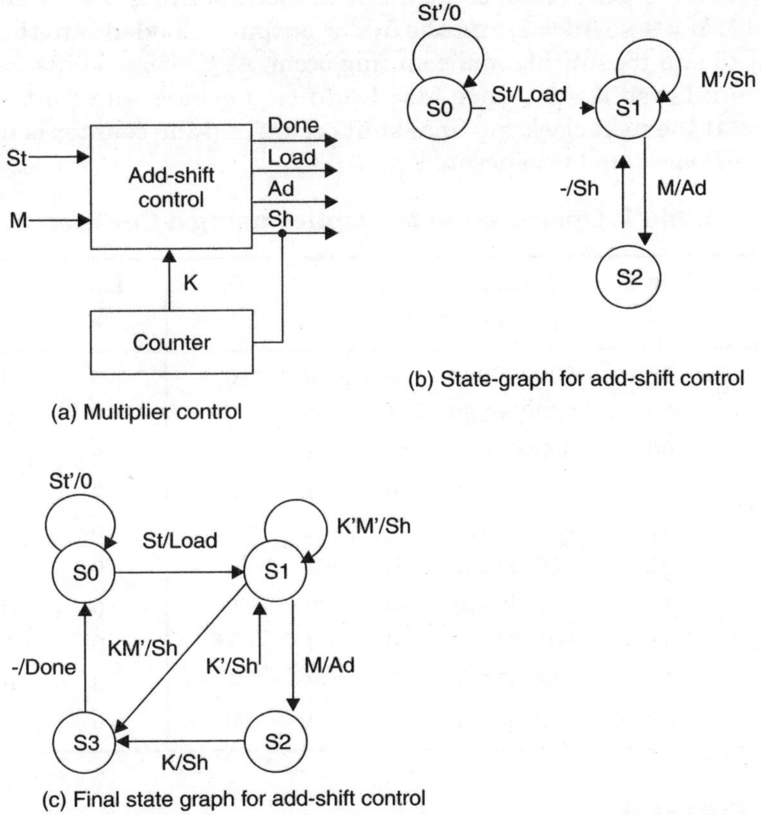

(a) Multiplier control

(b) State-graph for add-shift control

(c) Final state graph for add-shift control

Fig. 16. Multiplier Control with Counter

In order to determine when the multiplication is completed, the counter is incremented each time a shift signal is generated. If the multiplier is *n* bits, *n* shifts are required. We will design the counter so that a completion signal (K) is generated after *n* – 1 shifts have occurred. When K = 1, the network should perform one more addition id necessary and then do the final shift. The control operation in figure 16(*c*) is the same as figure 16(*b*) as long as K = 0. In state S1, if K = 1, we test M as usual. If M = 0, we output the final shift signal and go to the done state

(S3); however if M = 1, we add before shifting and go to state S2. In state S2, if K = 1, we output one more shift signal and then go to S3. The last shift signal will increment the counter to 0 at the same time the add-shift control goes to the done state. As an example, consider the multiplier of figure 14, but replace the control network with figure 16(a). Since n = 4, a 2-bit counter is needed to count the 4 shifts, and K = 1 when the counter is in state 3(112). Table 1 shows the operation of the multiplier when 1101 is multiplied by 1011. S0, S1, S2 and S3 represent states of the control circuit (figure 16(c)). The contents of the product register at each step is the same as multiplication steps written in previous section.

At time t0 the control is reset and waiting for start signal. At time t1, the start signal St = 1, and a load signal is generated. At time t2, M = 1, so an Ad signal is generated. When the next clock occurs, the output of the adder is loaded into the accumulator and the control goes to S2. At t3, an Sh signal is generated, so at the next clock shifting occurs and the counter is incremented. At t4, M = 1 so Ad = 1, and the Adder output is loaded into the accumulator at the next clock. At t5 and t6, shifting and counting occur. At t7, three shifts have occurred and the counter state is 11, so K = 1. Since M = 1 addition occurs and control goes to S2. At t8, Sh = K = 1, so at the next clock the final shift occurs and the counter is incremented back to state 00. At t9, a *Done* signal is generated.

Table 1. Operation of Multiplier using a Counter

Time	State	Counter	Product Register	St	M	K	Load	Ad	Sh	Done
t_0	S0	00	000000000	0	0	0	0	0	0	0
t_1	S0	00	000000000	1	0	0	1	0	0	0
t_2	S1	00	000001011	0	1	0	0	1	0	0
t_3	S2	00	011011011	0	1	0	0	0	1	0
t_4	S1	01	001101101	0	1	0	0	1	0	0
t_5	S2	01	100111101	0	1	0	0	0	1	0
t_6	S1	10	010011110	0	0	0	0	0	1	0
t_7	S1	11	001001111	0	1	1	0	1	0	0
t_8	S2	11	100011111	0	1	1	0	0	1	0
t_9	S3	00	010001111	0	1	0	0	0	0	1

ARRAY MULTIPLIER

We want to design a multiplier that consists of an array of AND gates and adders. This multiplier will have an interative structure with no sequential logic or registers required. Table 2 illustrates multiplication of two 4-bit unsigned numbers, X3 X2 X1 X0 times Y3 Y2 Y1 Y0.

Each of the XiYj product bits can be generated by an AND gate. Each partial product can be added to the previous sum of partial products using a row of adders. The sum of output of the first row of adders, which adds the first two partial products, is S13 S12 S11 S10, and the carry output is C13 C12 C11 C10. Similar results occur for the other two rows of adders. (We have used to notation Sij and Cij to represent the sums and carries from the ith row of adders.)

Figure 17 shows the corresponding array of AND gates and adders. If an adder has three inputs, a full adder (FA) is used, but if an adder has only two inputs, a half adder (HA) is used. A half adder is the same as full adder with one of the inputs set to 0. This multiplier requires 16 AND gates, 8 full adders, and 4 half-adders. After the X and Y inputs have been applied, the carry must propagate along each row of cells, and the sum must propagate from row to row. The time required to complete the multiplication depends primarily on the propagation delay in the adders. The longest path from input to output goes through 8 adders. If t_{ad} is the worst-case (longest possible) delay through an adder, and t8 is the longest AND gate delay, then the worst—case time to complete the multiplication is $8\ t_{ad} + t_8$.

Table 2. 4-bit Multiplier Partial Products

				X_3	X_2	X_1	X_0	Multiplicand
				Y_3	Y_2	Y_1	Y_0	Multiplier
				X_3Y_0	X_2Y_0	X_1Y_0	X_0Y_0	Partial product 0
			X_3Y_1	X_2Y_1	X_1Y_1	X_0Y_1		Partial Product 1
			C_{12}	C_{11}	C_{10}			1st row carries
		C_{13}	S_{13}	S_{12}	S_{11}	S_{10}		1st row sums
		X_3Y_2	X_2Y_2	X_1Y_2	X_0Y_2			Partial Product 2
		C_{22}	C_{21}	C_{20}				2nd row carries
	C_{23}	S_{23}	S_{22}	S_{21}	S_{20}			2nd row sums
	X_3Y_3	X_2Y_3	X_1Y_3	X_0Y_3				partial product 3
	C_{32}	C_{31}	C_{30}					3rd row carries
C_{33}	S_{33}	S_{32}	S_{31}	S_{30}				3rd row sums
P_7	P_6	P_5	P_4	P_3	P_2	P_1	P_0	Final Product

In general, an *n*-bit-by-*n*-bit array multiplier would require n^2 AND gates, $n(n - 2)$ full adders, and n half-adders. So the number of components required increases quadratically. For the serial—parallel multiplier previously designed, the amount of hardware required in addition to the control circuit increases linearly with *n*.

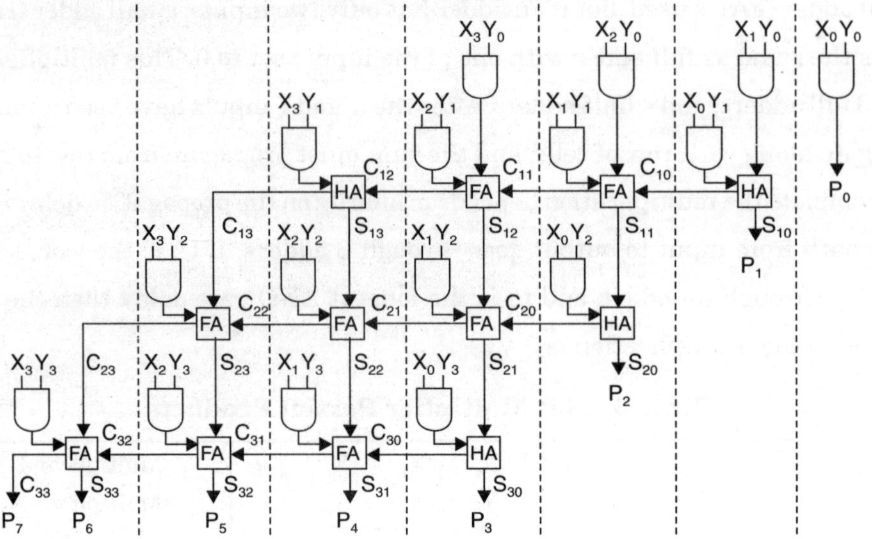

Fig. 17. Block diagram of 4 × 4 Array Multiplier

For an $n \times n$ array multiplier, the longest path from input to output goes through $2n$ adders, and the corresponding worst-case multiply time is $2n\,t_{ad} + t_g$. The serial-parallel multiplier of the type previously designed requires $2n$ clocks to complete the multiply in the worst case, although this can be reduced to n clocks using a technique discussed in the next section. The minimum clock period depends on the propagation delay through the n-bit adder as well as the propagation delay and setup time for the accumulator flip-flops.

Index

T

X